HISTORY OF BROADCASTING: Radio to Television

ADVISORY EDITOR

Dr. Christopher Sterling, Temple University

EDITORIAL BOARD

Dr. Marvin R. Bensman, Memphis State University
Dr. Joseph Berman, University of Kentucky
Dr. John M. Kittross, Temple University

The Outlook for Television

ORRIN E. DUNLAP, JR.

ARNO PRESS and THE NEW YORK TIMES
New York • 1971

Reprint Edition 1971 by Arno Press Inc.

© 1932 by Orrin E. Dunlap, Jr.
Reprinted by arrangement with Harper & Row, Publishers, Inc.
All rights reserved

Reprinted from a copy in The Newark Public Library

LC# 73-161143
ISBN 0-405-03564-0

HISTORY OF BROADCASTING: RADIO TO TELEVISION
ISBN for complete set: 0-405-03555-1
See last pages of this volume for titles.

Manufactured in the United States of America

THE OUTLOOK FOR
TELEVISION

To Mr Orrin E Dunlap. Jr
Very Cordially
Guglielmo Marconi

THE OUTLOOK FOR TELEVISION

BY
ORRIN E. DUNLAP, Jr., B.S.
*Radio Editor, "The New York Times";
Author, "Dunlap's Radio Manual,"
"The Story of Radio," "Advertising
by Radio," "Radio in Advertising";
Member Institute of Radio
Engineers*

INTRODUCTION BY
JOHN HAYS HAMMOND, Jr.
President, Radio Engineering Company of New York, Inc.

FOREWORD BY
WILLIAM S. PALEY
President, Columbia Broadcasting System

HARPER & BROTHERS PUBLISHERS
NEW YORK AND LONDON
1932

THE OUTLOOK FOR TELEVISION
COPYRIGHT, 1932, BY ORRIN E. DUNLAP, JR.
PRINTED IN THE U. S. A.
FIRST EDITION
E-G

To
HELEN and ARTHUR

CONTENTS

PREFACE ix
INTRODUCTION BY JOHN HAYS HAMMOND, JR. xi
FOREWORD BY WILLIAM S. PALEY xiii

Part I. Television—The Great Kaleidoscope

I. THE LONG ROAD INVENTORS TROD 3
II. RADIO IS GIVEN EYES 18
III. AN ELECTRIC HEART THROBS IN SPACE 31

Part II. The March of Television Begins

IV. EXPERTS ANALYZE THE PROBLEM 47
V. LIFE IS INSTILLED IN THE IMAGES 65

Part III. A New Decade in Radio Vision

VI. SEEING ACROSS THE ATLANTIC 79
VII. TELEVISION IN NATURAL HUES 91
VIII. FACES ON WIRES—FACES IN SPACE 106
IX. A CASTLE AND CITY OF DREAMS 127
X. VAULTING ACROSS TEN YEARS 145

Part IV. The Calendar Turns Again

XI. TELEVISION TECHNIQUE AND ARTISTRY 159
XII. TINY WAVES THAT SEE 176
XIII. A FLYING SPOT OF MAGIC 192

Part V. A Glimpse Ahead

XIV. TELEVISION'S COMMERCIAL DESTINY 221
XV. FACES AND SCENES ADRIFT 251

Epilogue

The possible effect of television on various fields of human activity is discussed by

BRUCE BARTON, *advertising and publications* 260

CONTENTS

REAR ADMIRAL RICHARD E. BYRD, *exploration*	260
DR. GEORGE B. CUTTEN, *education*	261
DR. LEE DE FOREST, *home and the theater*	262
BISHOP JAMES E. FREEMAN, *religion*	263
MAJOR-GENERAL JAMES G. HARBORD, *war*	263
COLONEL THEODORE ROOSEVELT, *politics*	264
S. L. ROTHAFEL (ROXY), *the stage and screen*	265

Appendix

CALENDAR OF WIRELESS-RADIO-TELEVISION	266
TELEVISION STATIONS IN UNITED STATES AND CANADA	288
INDEX	291

PREFACE

The purpose of this book is to reveal the romance of television and its commercial possibilities; to record historically the evolution of a new era in radio science; and to explain its magic. Today it is clenched in the hands of fate, ready to be freed and unfettered for a flight through space to entertain and to educate mankind.

Television is traced here step by step in chronological order, with dates accurately listed, so that no important advance in its development will be neglected or lost. It is necessary to go back a bit into radio to pick up the scientific threads from which the television pattern is woven.

The idea of seeing across wires and through space is not new. Television has been envisioned for years. Its historic record discloses the long route man travels to discover new scientific theories and truths. It reveals how quick man is to discard the old and grasp new ideas that carry him toward his goal. On these pages are found descriptions of various electrical systems, devices, and ideas devoted to the search for television, that man may eventually see to the ends of the earth.

The author has been privileged to discuss the science of seeing by radio with Marconi, deForest, Alexanderson, Ives, John Baird, and other contemporaries in television research. Their outstanding characteristic is modesty, a quiet, unexcited nature. They speak softly. They are not publicity-seekers. They do not boast. They all realize that television is just beginning and that they are faced with numerous problems, some of which another generation will be called upon to solve. They refrain from whimsical prophecy. When some fantastic use of television is suggested, they smile and throw up their arms, more amazed than the layman. Their task is to make television practical.

The inventions here mentioned, the men whose ideas point the way, stand as the beacon lights of television. What future generations achieve will tower on the bedrock that the men of yesterday and today leave behind. The images and scenes that girdle the earth even one hundred years hence will have in the power of their wings the hours of toil, the disappointments, and the triumphs of the men who breathed life into television from 1880 to 1933.

NEW YORK CITY.

O. E. D., JR.

INTRODUCTION

In radio the gold rush is over. The incredible growth of this infant of industries brought into its onrush all those opportunists who are ready at a moment's notice to become anything that will bring them money. Radio engineers developed overnight, sales experts talked glibly of circuits, and attics were transformed in a few days to manufactories for radio equipment. The depression will, among the many good things that it is contributing, probably put radio on a basis of sounder economics.

It is my belief that this industry will cast off those who gamble on its future and will retain those who revere it as a splendid science and who are willing to serve it in a truly scientific sense. Much engineering lies ahead of us; the major problems of fading and static are as vital today as twenty years ago. The limitations of the natural medium through which our stations operate are becoming daily more definitely understood. Efficiency in our engineering results is our goal. While much of the poetry of radio is disappearing, none of its romance has been lost.

I often feel that it is dangerous to prophesy too definitely on the technical future of a science. The obvious seldom happens, and so often collateral developments take place around an industry which are outgrowths of it, and which at times grow into greater magnitude than the parent industry itself.

The great field of electronics is leading the vacuum tube into a hundred uses which none of us could foresee a decade ago. It is a far cry from the reception of radio signals to the segregation of coffee beans, but the vacuum tube has in its versatility become the key mechanism in both operations.

Undoubtedly, the functions we now see taking place in

the technique of radio apparatus will lead into vast fields of human endeavor, and even the imagination of the inventor can only arrive at a vision of these new fields in a slow, laborious manner. The coordination of human knowledge, the constant overlapping of problems in diversified industries, and the closer engineering contacts now established between scientists and engineers in different fields will produce new sciences in ever increasing numbers.

Inventing is getting to be easier. We have at hand so many new facts to handle and to reassemble. It is possible to move ahead more swiftly now, for many of the snags have been removed by the technical experience of others.

There is also the growing freemasonry of sympathetic help in one science toward another. Many of the theological and social aims are achieved daily and unconsciously by those who are cooperating in the cause of science and the fight for the truth.

Today, television is opening its eyes!

JOHN HAYS HAMMOND, JR.

FOREWORD

While the Morses, Fultons, and Watts were greeted with incredulity, even open resentment, when the secrets they had locked within their laboratories were bared to the public, television is being born into a new and different world. Today the public is demanding "When" even more vociferously than it asks "How."

Predictions that visual broadcasting will serve as some genii opening up new worlds of culture and contentment, outstrip the achievements of today, even if they do not go beyond the possibilities which are apparent to all of us who have had some practical contact with this new application of science and art.

We are asked insistently every day, when will visual transmission develop to a state of perfection with definite and clear images flashing through the ether into the home? We are quizzed as to when will the television receiving-set be able to etch vividly the tense play of a football game on a distant field? When will radio-without-sight become as obsolete as motion-pictures-without-sound?

Mr. Dunlap's book is published as these questions become more numerous and may do much to indicate the answer to them. For my own part, I find prophecy impossible. I believe, in view of the many good minds working on television's problems and in view of the elaborate facilities made available for modern-day research, that the perfection of television is as certain as tomorrow's sunrise. But to predict its perfection is one thing and to prophesy its applications is another.

The possibilities are so varied and the implications so many-sided that it is impossible to forecast even a small portion of the uses of visual broadcasting. Enough to say, it should bring a new era in educational entertainment, should become the greatest force for education ever developed. Its perfection should mean the development of new forms of dramatic and musical presentation—works planned to fit a new medium and developed to take full advantage of its greater freedom.

But to name a date is to get ahead of the technicians and scientists who are working day and night to resolve the riddle of sight transmission. When they have finished their lonely job of scientific pioneering, there will be many ready to apply those principles to everyday life.

WILLIAM S. PALEY.

Part I

TELEVISION—THE GREAT KALEIDOSCOPE

CHAPTER ONE

THE LONG ROAD INVENTORS TROD

THE world is on the threshold of a great forward movement in mass communication—transmission and reception of sound and sight combined. Just as the incandescent lamp guided man out of the dark ages and the motor-car extended his highways, so radio came to introduce a kind of armchair civilization. The snap of a switch brings music, drama, and speech delivered wholesale to a multitude who tap the electrical flow on a slender copper wire. The science of broadcasting presents the world with new instruments for reaching a populace of many millions. And so the machine age has ushered in an era of electrical entertainment, the next act of which is television! It is a gigantic force looming to take its place beside the press, the talking picture, and radio broadcasting as a powerful instrumentality for entertainment and education.

Television is a science and an art endowed with incalculable possibilities and countless opportunities. It will enable a large part of the earth's inhabitants to see and to hear one another without leaving their homes. Sight on the wings of radio is an easy, quick, and economical means of spreading knowledge and information. Eventually it will bring nations face to face, and make the globe more than a mere whispering-gallery. Radio vision is a new weapon against hatred and fear, suspicion and hostility.

Revolutions in modes of communication have always made the world more interesting. The word of mouth, the scratch of a pen, the metallic clicks of the telegraph, the spoken words of the telephone, the buzzing dots and dashes of wireless, the music and voices of broadcasting, all worked vast changes for individuals and institutions. And now transmission of sight by invisible waves in space is destined to stretch man's horizon and to stimulate new interest among a great variety of people and things.

The curtain is rising on a new era, a sort of educational renaissance, far removed from the age of exploration and adventure that beckoned man into new environment, tested his courage and called for strenuous physical endeavor to carve villages out of the wilderness. A new wizardry is evolving in which the eye is restricted no more than the ear.

Television is the science of seeing by radio or wire. Electrically it prolongs the optic nerve empowering the human eye to scan a distant scene or person. *Tele* is Greek for "at a distance." *Video* is the Latin verb meaning "I see." Thus the Cæsars might have said, *televideo*—"I see at a distance."

Television in reality is a second-sight, officially defined as the electrical transmission of a succession of images and their reception in such a way as to give a substantially continuous and simultaneous reproduction of the object or scene before the eyes of a remote observer.

The wonder-working eye that does the trick in this enchanted science is the photoelectric cell. Photo is derived from the Greek word *phos*, meaning light, while electric is from the Greek *elektron* and the Latin *electrum* meaning amber. Thales of Miletus was first to notice that when amber is rubbed it becomes capable of attracting light bodies such as bits of paper or straw; that was the first electrical phenomenon produced by man.

This photoelectric "eye" is sensitive to light. It represents the combined action of light and electricity. When properly wired and subjected to illumination it transforms lights and shadows into corresponding electrical impulses, which flow through the air as electromagnetic waves or along the wires as electricity.

The next touch of magic at the receiving station is to convert the waves back into electricity and then into light without losing the identity of the originally televised scene. The photoelectric cell in its performance does for light what the microphone does for sound. The microphone is radio's ear; the photoelectric cell is the eye.

Television has not arrived over any short highway of science. There have been no short cuts. It has been a long, tedious journey which began almost a hundred years ago. The road has been strewn with obstacles. And, while some, who lived in the mauve decade and beyond, dreamed of seeing across wires or through space, they were none too sure that it might be practical. The experimenters encountered many barriers of discouragement. It was not until about 1920 that man was convinced that he had within grasp the perfected instruments necessary for broadcasting voices, music, and sundry sounds. He then began to realize that the next logical step would be to combine sound and sight in broadcasting.

In the haste of this day one is likely to award all the laurels to the research workers of the present age for developing television. Before doing this, however, some tribute should be given the tireless workers who years ago had the foresight to attempt transmission of pictures by wire and later by wireless. The pioneers were handicapped. They lacked the devices, which only time could bring forth, to make their dreams come true. They erected signs along the highway of scientific knowledge, pointing toward the goal of success.

Those Who Share the Honors.—Television is not the triumph of any one man, but of many. No one can be called the inventor. Naturally, the first experiments in transmission of pictures were by wire. Radio was too elusive. There were no vacuum tubes for amplification, no sensitive electric eye, in fact, no wireless, in the days when Alexander Bain, of London, in 1842, arranged metal letters on a conducting plate at a sending station and a chemically prepared paper on a similar plate at the receiving end of the line. Narrow conducting brushes, mounted side by side on an insulated strip, were moved slowly over their respective plates. The brushes at corresponding positions at the two stations were connected by individual wires so that when contact was made with the metal letters, current passed through the paper, producing a discoloration in the form of the letters at the sending terminal. Obviously this system required far too many wires.

F. C. Bakewell, of England, in 1847, using two metal cylinders driven at the same speed, transmitted a graph drawn with an insulating ink of shellac on one drum for interrupting the current to a chemical paper on the other. A single brush in contact with its rotating drum was given a slow longitudinal motion, causing it to trace a spiral on the drum, thereby transversing the entire area of the graph. Only one wire was needed between the stations, with a ground return. Man is indebted to him for two fundamental ideas, namely, the use of rotating cylinders and the longitudinal motion of the exploring element, both of which are still widely used in photoradio and in transmission of pictures by wire.

A man named May, an operator in the Atlantic cable station at Valencia, Ireland, observed by chance that light shining through a window on some selenium resistance units unbalanced his bridge circuit. A few years later, in 1873, Willoughby Smith, employing selenium as a resistance, was annoyed by its instability. He investigated the source of the variations and made the important discovery that its resistance decreased with the intensity of the light falling upon it. After that selenium was used by many of the pioneer experimenters in picture transmission.

It was Philip Reis, of Germany, who advanced the theory that light falling upon selenium liberates electrons which assist in conducting the current. An investigation by Elster and Geitel, in 1889, disclosed that various elements possess photoelectric properties—that is, they emit electrons when illuminated. Among these metals are thalium, strontium, lithium, sodium, potassium, rubidium and caesium.

Stoletow, in 1890, made the first photoelectric cell, using zinc for the cathode, which necessitated ultra-violet light as an exciting agent. Electrons were attracted to a platinum plate which was made positive with respect to the zinc by a high-voltage battery. The photoelectric cell was further developed to its present state of sensitivity by the proper handling of these electro-positive, chemically active metals in a vacuum.

Despite the fact that he lacked the present-day devices and ideas, N. S. Amstutz, an American, in 1890, is credited with having sent the first successful picture with half-tone over a twenty-five-mile wire line in eight minutes.

Professor Arthur Korn, in Germany, about 1902, wrapped a photographic negative around a glass cylinder which was rotated and at the same time moved along its axis so that light from a point source traversed every portion of the negative. The amount of light passing through it on a selenium cell varied with the density of the negative, thereby producing a variation in the line current transmitted to a distant station. Korn sent a picture of President Fallières of France from Berlin to Paris by wire in twelve minutes in 1907.

A serious objection to selenium was its slow response to rapid variations of light intensities. History records, however, that the results obtained by Amstutz, Witherspoon, Korn, Ruhmer, Belin, Leishman, and others were of fair quality. Little did these pioneers realize, when experimenting with picture transmission and electromagnetic waves, that some of their observations and discoveries would eventually lead to a worldwide communication system and to an international sound-sight theater of the air.

HERTZ PRODUCES THE WAVES.—Heinrich Hertz, of Karlsruhe, Germany, confirmed Clerk Maxwell's theory of ether waves in 1886, by creating and detecting electromagnetic waves. Incidentally, that is why radio waves are sometimes called Hertzian waves. Hertz also discovered in his experiments with wireless waves that ultra-violet light falling on a spark gap permitted an electric discharge to take place more readily than when the gap was in darkness.

An allied effect was uncovered the next year when Wilhelm Hallswach, a German physicist, noticed a well-insulated and negatively charged body lost its charge when illuminated with ultra-violet light. Today man can look back and see in those observations highly important theoretical beginnings in the evolution of television. But in 1888 those strange electrical effects were too feeble to suggest anything of practical significance.

A serious handicap encountered by early experimenters was lack of electrical amplification. Invention of the two-element valve by John Ambrose Fleming, in 1904, cast some hope in this direction, although it was a detector rather than an amplifier. Finally, in 1906, a remarkable advance was made when Lee deForest, invented the three-element vacuum tube, which he named the audion. It gave the weak currents renewed strength. It was exactly the device which telephony and radio had long been awaiting. The vacuum tube gives electricity from the photoelectric cell real power; consequently, what once appeared to be trivial sparks and minute electrical impulses, have surged into a powerful radio force that gives wings to sound and sight. The audion strengthened television's eyes.

PHOTORADIO LEADS THE WAY.—It is natural that the first step in the evolution of television should have been wireless, under which name the medium developed. Then came

THE LONG ROAD INVENTORS TROD

sound broadcasting as an advance from the dots and dashes. Facsimile transmission was next to be followed by motion pictures in the air, just as the stereopticon pointed the way to the silent cinema and the talkies.

Captain Richard Ranger contributed to photoradio or picturegrams. It was on May 7, 1925, his invention was used to send war game pictures and maps 5,136 miles in 20 minutes from New York to Honolulu. A photographic film revolved on a glass cylinder over which played a powerful needle or pencil of light. The black detail of the picture checked the light passage and the lighter areas let it get through. This light of varying intensity fell upon a photoelectric cell which transformed the light into electrical impulses so controlled that the pattern of the original picture was preserved at the distant receiving point. Briefly, the picture was first traced in light. The light was converted into electrical current. The current was amplified a few million times and broadcast. At the receiver the radio signal was intercepted and again converted into electrical current, which operated a pencil of light that resketched the picture on a paper wound around a cylinder revolving in step with the one at the transmitter.

The Ranger apparatus was utilized on April 20, 1926, to send a picturegram of a $1,000 check from London to New York where it was cashed by the Bankers Trust Company. It was signed by Major General James G. Harbord.

On June 11, 1927, pictures were radioed from London and from Hawaii to the Massachusetts Institute of Technology dinner in New York by means of an improved Ranger system. In this demonstration a stream of hot air driven through a small bore muzzle moved to and fro across a special paper treated with nickel. The incoming radio signals operated the hot-air gun, the heat from which turned the paper on the revolving cylinder from white to sepia wherever its pencil-like line traced. In this way it reconstructed line by line pictures flashed by facsimile radio from stations across the sea.

AMONG THE CONTEMPORARIES.—Year after year many scientists, assisted by a corps of research experts, have come forward to add their magic touch and knowledge to the progress of television. The advance has been slow and at times discouraging. But they plodded on. At last their fame began to grow rapidly, after broadcasting got under way in 1920. The achievement in perfecting the sound broadcasts gave them a strong foundation and bridged several gaps enabling them to concentrate on the development of seeing by radio.

Prominent among the contemporaries who have devoted their time and energies to television are: Dr. Ernst Frederik Werner Alexanderson, Dr. Herbert E. Ives, John Logie Baird, Vladimir Zworykin, John Hays Hammond, Jr., Dr. August Karolus, Philo T. Farnsworth, Ulisses A. Sanabria, R. D. Kell, development engineer, who assisted Alexanderson and Zworykin, C. Francis Jenkins, and a host of workers who have faithfully toiled with these men to rush the perfection of instruments that enable their fellow men to see by radio.

Hans Knudsen is credited with sending a photograph by wireless in 1909. And in 1910, A. Ekstrom, a Swedish inventor, discovered he could scan an object directly by a strong light beam. He placed the source of light on one side of a scanning disk and the light-sensitive element or "eye" on the other side.

BACK IN 1884.—Paul Nipkow of Germany, a farsighted youth, has the distinction of inventing the spirally perforated scanning disk which was indispensable in early television. He introduced the device in 1884. His idea was to dissect pictures using a light-sensitive cell, a lens and a scanning disk. This was the underlying principle of the first television systems.

The trick has always been to discover more scientific methods of this form of radio surgery and to secure the benefits of intense illumination. The scene or image must be cut up into tiny fragments for broadcasting. At the receiving end the pieces are plucked from space and woven into a duplicate of the original picture. Each line must be painted electrically in proper sequence else the identity is lost. Thousands and thousands of dots of light flit across the screen, but the eye is not fast enough to see them all. It catches sight of only the complete picture.

JENKINS BEGINS TO SEARCH.—Nipkow ran into many obstacles. The light-sensitive cells of that time were not fast enough to reproduce images in motion. He lacked the neon lamp, the cathode ray tube and the photoelectric cells which helped others in later years.

C. Francis Jenkins, in 1890, began the search for new appliances that Nipkow's disk needed for success. He began by dropping pennies in a slot machine and watching the strange, animated images. At the Atlanta Cotton Exposition in 1895, he demonstrated motion pictures. Then the idea occurred to him to send pictures by wireless.

He was a pioneer in attracting the American public's attention to television. As early as 1922 he predicted motion pictures by radio in the home, and that "entire opera may some day be shown in the home without hindrance of muddy roads." In 1923, he placed a portrait of President Harding in a camera-like affair at the Naval Radio Station in Washington and it was plucked from the air 130 miles away, atop the *Evening Bulletin* Building in Philadelphia. Two years later Jenkins predicted that miniature motion picture screens would some day be attached to radio sets in every household. He conducted experiments along this line and announced that he expected "to stage a radio-vision show with the talent performing at the broadcasting station and the audience watching at the receiving station miles away."

He invited government officials to watch the blades of his windmill-like machine casting images, blurred but nevertheless distinguishable on a screen.

OTHERS TAKE UP THE WORK.—Dr. Ernst F. W. Alexanderson, in "the House of Magic" at Schenectady, directed intense research in radio vision, in sending and receiving images and in wave propagation. His fame spread, when in 1930 he showed television pictures on a theater screen six by seven feet.

Dr. Herbert E. Ives, electro-optical research director of the Bell Telephone Laboratories, was the first to show a radio camera that would televise outdoor scenes without the glare of artificial lights. Later he demonstrated television in color, and next two-way television in which the speakers at the ends of a telephone line saw each other as they conversed.

John L. Baird thrilled London with radio vision. He developed the instruments that enabled the officer of an ocean liner to see a pretty girl on land 1,000 miles away. He sent a face across the sea and later televised the English Derby.

It was John Hays Hammond, Jr., who in 1930 revealed he had patented an electrical system equipped with a television eye to aid aircraft in landing at airports, no matter how thick the fog or how inky black the night.

Vladimir Zworykin won recognition by developing a television receiver which utilized the cathode ray tube, and thus dispensed with the scanning disk and other movable parts. He simplified the apparatus and made it more commercially practical for home use. He scanned the object electrically instead of mechanically.

Philo Farnsworth did likewise in his California laboratory and then came east to Philadelphia to develop his receiving machine. He said the system he employed required only a narrow pathway in the radio spectrum.

Dr. August Karolus perfected an electro-chemical light valve which facilitated more powerful illumination of the object to be televised. It controlled the flow of light with great rapidity. Up to this time all mechanical shutters had failed to operate at sufficient speed. The Karolus valve helped to push television ahead.

Hollis Baird (no relation to John L. Baird of Scotland) conducted television experiments with mechanical scanning at Boston. He developed a scanner in the form of a horizontal metal plate called a "spider," which supported a narrow strip of thin steel perforated with square holes.

Ulisses A. Sanabria, in April, 1931, showed television on a two-foot screen in his Chicago laboratory. The close-ups were described as "marvelous." Sanabria used what he called a "lens disk," a solid aluminum wheel with forty-five lenses sunk into it. A zipping daub of light caused by the disk revolving at a high speed flooded the screen with light. He tinted some of the faces by employing a neon-mercury gas in a special lamp designed by Warren B. Taylor. Later Sanabria demonstrated images on a ten-foot screen at the 1931 Radio-Electrical World's Fair in New York.

Thousands went to the exhibition to see the images. The man in the street began to ask "What is this thing called television?"

KENNELLY's DEFINITION.—And for those who want to know, Arthur E. Kennelly, professor of electrical engineering at Harvard University, defined television and explained the process in the *Annals* of the American Academy of Political and Social Science as "the instantaneous transmission to a distance, of the image of an object, so that the persons at the receiver can see the reproduced image and thus in a certain sense see the object itself.

"There is a crude resemblance between the principle of television and that of telephotography," said Kennelly. "In both there is a rotating pair of similar elements running in close synchronism, so that corresponding points in the sent and received pictures are simultaneously projected. Whereas, however, the photographic films in telephotography may take several minutes to execute from beginning to end, in the case of television, the two pictures must be completely covered in about one sixteenth of a second, in order that the eye may see the whole surface as a single picture.

"In one form of the apparatus, a bright beam of light is caused to travel in a definitely repeated manner, over the object to be televiewed, with the aid of a series of holes in a rapidly revolving disk. The light, reflected from successive areas of the object, is directed to a photoelectric cell, in such a manner that bright spots on the object simulate strong currents in the cell, and dark spots feeble currents. These currents, greatly amplified, are delivered to the air at the sending mast.

"A minute fraction of the emitted wave energy is picked up at the receiving mast and delivered, after reamplification, to control the instantaneous intensity of a beam of light from a local source, directed through holes in the receiving disk, to corresponding parts of the received picture. The bright and dark spots of the sending picture will then reappear as corresponding bright and dark parts of the received picture. In this way, several thousand successive points in the sending picture will, one by one, be reproduced in the received picture, all run over sixteen times per second. Changes in the form and brightness of the object will simultaneously appear in the reproduced image at the receiving station."

MARCONI'S CONQUEST.—Few fathers have been able to predict the destiny of their child with the accuracy that Guglielmo Marconi has been able to foresee the steps that his wireless would take from year to year. He has always looked ahead to the day when wireless vision would be a reality.

Early in 1931 he was asked how soon he thought television would be practical.

"Television is the highest grade in the art of communication and it is rapidly benefiting from the improvements made in the lower grades of the art-telephonic and rapid picture transmission," said the inventor. "I think that when the latter are further perfected we may be close to practical television."

The history of wireless leads on to television. Although Marconi in the nineties did not attempt to send pictures, he pioneered in television when sending dots and dashes across his father's estate at Bologna, Italy. The men who chopped the pathway through the forests for the first transcontinental railroad did just as much for the advance in transportation as those who laid the track. Marconi by discovering how to utilize wireless waves for communication cut the pathway through the sky over which the television images of the twentieth century could travel from city to city and from nation to nation.

"What that means for mankind no one can even guess," said Sir George R. Parkin, a professor at Upper Canada College, after he saw Marconi send the first west-east wireless message from Glace Bay in 1902. "The path to complete success may still be long and difficult. Between George Stephenson's 'Puffing Billy' and the great mogul engine which swings the limited express across the American continent, there lies three quarters of a century of endeavor, experiment and invention. But in the great original idea lay the essential thing which has revolutionized the world and the conditions of transportation. I came away from Glace Bay with the feeling that Mr. Marconi's modest confidence in his work will in the end be justified by results. Meanwhile, patience may still be necessary. Weeks, months, even years may be required to bring the system to complete efficiency."

THE INVENTOR RECOLLECTS.—Twenty-five years after this triumph at Glace Bay, Marconi called the spanning of long distances by radio child's play compared with the uncertain task in 1902.

"A quarter century ago the instruments we had at our disposal were very crude compared with those we have today," said the inventor. "We had no vacuum tubes, no sensitive superheterodynes, no amplifiers, no directional transmitters and receivers, and no way of making continuous waves. All we had for transmitting was the means of making crude damped spark waves, which did not permit the accurate tuning we have today.

"As to the application of wireless in the future I am always averse from entering into the realm of prophecy, but perhaps I might suggest that, apart from the ordinary transmission and reception of wireless messages there is a possibility that the transmission of power over moderate distances may be developed, and that television will become an actuality. I must leave to your imagination the uses which can be made of these new powers. They will probably be as wonderful as anything of which we have had experience so far.

"Looking back at our old difficulties, the ease and perfection recently achieved by radio, especially in regard to broadcasting, appears little short of miraculous. It shows us what can be done by the combination of a great number of workers all intent on securing improved results. And how many, who began as amateurs, have contributed in one form or another to the progress and success?

"We are yet, however, in my opinion a very long way from being able to utilize electric waves to anything like their full extent, but we are learning gradually how to use the wireless waves and how to utilize space, and thereby humanity has attained a new force, a new weapon which knows no frontiers, a new method for which space is no obstacle, a force destined to promote peace by enabling us

THE LONG ROAD INVENTORS TROD 17

better to fulfill what has always been essentially a human need—that of communication with one another."

Thus television has been launched. Its commercial and æsthetic possibilities are seen as tremendous. Research experts are at work refining the instruments, clarifying the images, enlarging the pictures on the screen and preparing for the inevitable welcome at the firesides of all nations—but much work remains to be done.

Chapter Two

RADIO IS GIVEN EYES

When man discovered that he could send dots and dashes through space without the use of interconnecting wires he called it wireless. The next logical step was to extend the range of the voice by radio, and that was called radio telephony. Then the ether, or whatever that mysterious medium is that occupies all space, was caused to vibrate with music and entertainment. This new magic that entertains millions of listeners simultaneously in their homes was called broadcasting. The next move was to broadcast sight. That is called television. And so the sound broadcasts reach the brain through the ears while radio vision is for the eyes—"those marvelous little mechanisms, which stand as twin entrances to the brain."

Light enables the eyes to see. Man cannot see behind him or around corners unless he uses mirrors or lenses to bend or reflect the light. The motorist sees what is on the highway behind him by means of a little mirror above the windshield. The sailor in the submarine glimpses above the surface of the sea through the spyglass of the periscope. The soldier scrutinizes the landscape with powerful binoculars. The astronomer with his wondrous telescope peers far out into the heavens to scan other worlds. The scientist sees the bacillus by use of the microscope. And now by television the range of the optics is given greater scope. Man can literally look through mountains, through thick walls and across the sea. Television removes barriers, which throughout the ages have restricted the range of the human eye.

Distance will not limit television in its ultimate form. Radio, the wings upon which the scenes travel, has skill to girdle the globe at the speed of light. It empowers man to talk around the world in the twinkling of an eye. It will enable him to see around the earth. Television ignores mist, smoke, clouds and darkness. It looks through the blackness of the night.

Just as radio brings the chirp of a canary, the buzz of a bee, the whisper of a child or the moan of a violin across the horizon to vibrate the eardrum, so does television prolong the optic nerve so that it may distinguish scenes and people in action far across the countryside.

BARRIERS ARE CONQUERED.—When three stars hang on a door in the Bell Telephone Laboratories they mean "positively no admittance." And those about the place know something important is going on inside. The stars glitter from time to time on the door of the television laboratory. That means a new theory has been found and the research experts are testing its practicability. And when theories are lacking a game of checkers may be in order. Checkers rest the mind and then new ideas are likely to crop out.

It was April, 1927, when the research engineers left their three-starred room of mystery to demonstrate what they could do with television between New York and Washington, D. C. They proved beyond a doubt that it is physically practical to make an extensible optic nerve although it is physiologically an impossibility.

On November 26, 1927, in an address before the Association of Science Teachers of the Middle States, John Mills, of the Bell Laboratories, proclaimed for the engineers that no longer did the eye of man require a free, clear, straight path to view a distant object, scene or person. Television is at a stage where it places the eye in a satisfactory position to view distant objects, because radio transmits observations through intervening barriers, which the eye without television cannot penetrate.

Those who listened to Mills at this meeting in Atlantic

City were told at the beginning that to understand television one must realize that the first lesson involves an elementary exposition of physics and chemistry as well as electricity and radio. In as non-technical a way as possible he revealed the wizardry of the research experts who had succeeded in prolonging the sensitive optic nerve that runs from the eye to the brain. And this is how it is done.

"An electric eye is placed before the object, which must be sufficiently illuminated in order to be observed in an electrical manner," said Mills. "To the location of the distant spectator there stretches an electrical circuit through which the electric eye transmits its observations. Figuratively speaking, this circuit acts as an extension to the optic nerve. Unlike an actual nerve channel it cannot terminate directly in the brain of the observer. Therefore, it terminates in certain electrical equipment—the viewing apparatus, which reproduces as a picture the scene viewed by the distant electric eye. The observer does not see the scene itself. What he sees is an image of the scene, its optical counterpart. Flashes of light, originated in the viewing apparatus by the action of the distant electric eye, create for his eyes a visible presentation of the scene. The effect is much the same as if he viewed a small and very bright screen-presentation of a motion picture or cinema reproduction of the distant scene.

AKIN TO TELEPHONY.—"In certain respects the apparatus and methods of television are like those of telephony. One is an aid to seeing, and the other to hearing. One reproduces remote scenes for an observer, and the other distant sounds for the listener. In telephony an 'electric ear' is placed near the source of the sound. This is the familiar transmitter, an electro-mechanical device which is sensitive to sound waves. By it the mechanical action of the sound waves is converted into an electrical effect. When sound waves impinge upon its diaphragm electrical currents arise; the motions of the electrons, the minute particles of electricity which constitute these currents, correspond and are similar to the motions of the molecules of air which constitute the sound waves. The telephone transmitter, in other words, is a sound-sensitive device which can give rise to an electrical current, corresponding in its variations to sound waves and thus embodying any speech significance they may have.

"At the other end of the telephone circuit is the telephone receiver, an 'electric mouth' which can utter sounds such as those of human speech. It is an electro-mechanical device, by the action of which electrical currents are converted into mechanical effects. When the current embodies the variations of a sound wave the diaphragm of the receiver vibrates and the adjacent molecules of air are forced into a corresponding wave motion. The telephone receiver is a sound-active device."

A CHANNEL IS ESSENTIAL.—There must be an intervening channel for passage of the electrical energy; and this path may be wire, radio or a combination of both. Radio transmitting instruments, on the other hand, comprise an electrical means for converting the energy of electrons, moving as a current in a wire, into electromagnetic waves that travel through space. It is the duty of the receiving set to reconvert the invisible wave motion into electrical current.

Thus, in television, similarly, the channel between the terminal apparatus may be entirely wire or part of it radio. So far as the passage of electrical energy is concerned, telephone and television systems are essentially the same. The terminal apparatus differ, of course, but are analogous.

Television requires a light-sensitive mechanism which acts as an eye, instead of a sound-sensitive device (microphone) which functions as the mouthpiece. And the distant observer needs a light-active device which originates light, instead of a sound-active mechanism (loudspeaker). The light-sensitive, electric eye converts the energy of light rays into electrical energy moving in wires; the light-active neon lamps reconvert these electrical currents into light.

NEW ELEMENT No. 87 MAY HELP.—The research experts are greatly interested in the discovery of element No. 87, which was found in the mineral samarskite by Professor Jacob Papish and Eugene Wainer of Cornell University, in October, 1931. It is said that "it will be similar to cæsium." And cæsium is used in television's eyes to make them sensitive to light fluctuations. This new element shows promise of greatly improving the sensitivity of the photoelectric cell, in fact, it has been estimated that a cell designed with No. 87 as the light-sensitive element could be placed on one side of a door an inch thick and it would be influenced by light on the other side. This being true, such a bulb would be an extremely sensitive electrical eye.

Discovery of element No. 87 fulfills a prophecy made by Mendeleeff sixty years ago. It has been provisionally known as "ekacæsium." Mendeleeff's remarkable table was published in 1870. At the top is hydrogen, the lightest of all elements, and at the bottom stands the heavy uranium as No. 92. Element No. 85 is still missing. Mendeleeff called it "ekaiodine." Chemists know in what group it lurks, but its isolation is a matter of skill and patience. Radium is element No. 88.

ALL MATTER IS ELECTRICAL.—In the behavior of the light-sensitive devices lies the great mystery of the matter of which the physical universe is composed. Scientific research has gone deeply into this during the past thirty years. The engineers in explaining the photoelectric cell find it necessary to touch a little on the constitution of matter.

All matter, it seems, is really electrical. All the eighty-eight, or so, different elements which the chemist knows, whether iron or iodine, calcium or carbon, exist in the form of atoms, small particles, invisible even to the most powerful microscope. These atoms in combination with each other form the molecules of all the myriad different materials which occur naturally in our world or have been produced by the ingenious chemist. Behind all this apparent complexity is an amazing simplicity. All the different atoms are alike in the substances of their composition. It is in the amount and arrangement of these substances that the atoms of different elements are unlike. Atoms are composed of two kinds of particles, known as electrons and protons.

NATURE OF RADIO EYES.—The photoelectric cell is a highly evacuated glass bulb coated on part of the inside with a light-sensitive material, which if properly prepared and exposed to light becomes electrically sensitive to illumination although it may be as feeble as a candle's glow. There are two wires leading into the cell. One connects to the light-sensitive substance on the wall of the bulb and the other to a ring of photoelectrically inactive metal such as nickel or platinum. Then, when light falls on the active surface electrons are emitted at a rate proportional to the quantity of light absorbed by the coating. These negative particles of electricity, free to move in the evacuated space, are attracted to a metal ring, the second electrode in the center of the bulb. A battery keeps this electrode positive. During the passage the electrons collide with molecules of argon, and since their velocity-voltage is higher than the ionizing potential of the argon, ionization occurs. Thus the electrons stream through the wires and into a measuring instrument which indicates a current. This flow of electricity is produced by action of the light; the energy of light, its ability to do work, is converted into electrical energy, into a motion of electrons, which in turn can do work.

No current can flow through the cell except as electrons are released from its photoelectric terminal by action of the light. As fast, however, as electrons are emitted they are drawn across to the collecting ring and through wires to the battery, while others leaving the negative terminal of the battery hasten to replace them. The process continues as long as the cell is exposed to light; and the electrons sweep around the circuit like a widely scattered field of riders in a six-day bicycle race, according to the Mills' description.

At any instant the number of electrons passing any point of the circuit is just the number at the same instant emitted from the photoelectric surface. The more intense the light the greater this emission. In fact, the current is always directly proportional to the light, and if that varies in intensity exactly corresponding variations occur in the current. That is why the photoelectric cell is employed as television's eye.

RECREATING THE SCENE.—"In a radio-vision system there must be complementary to the light-sensitive transmitter or electric eye, a light-active receiver, just as in telephony a receiver is complementary to the transmitter," Mills said. "This must give forth light in response to an electric current; and the intensity of the emitted light must be directly proportional to the current. Then, whatever light the electric eye sees may be recreated, and all the variations in the original illumination faithfully reproduced.

"For television it is necessary that the light-active device shall perform instantaneously in accord with its controlling current. An ordinary electric lamp-bulb would not serve because an appreciable time must elapse after the current is turned on before the filament heats enough to glow. And when the current ceases, the light itself does not stop at once but fades out gradually. An instance in nature, where light instantaneously accompanies the current which causes it, is found in the lightning flash. The same phenomenon, on a smaller scale and much controlled, is utilized in the design of the light-active element for a television system."

MINIATURE LIGHTNING FLASHES.—The layman and engineer are aware that enormous voltages are required in

THE ALEXANDERSONS AT HOME
The inventor shows the folks what the rest of the world will be doing
in years to come.

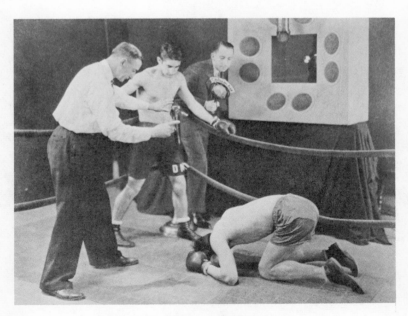

ELECTRIC EYES AT THE RINGSIDE
Boxing bouts—the clang of the gong, clamor of the crowd, and plenty of
action on the screen—are popular events on the air.

the atmosphere to accelerate electrons and the casual ions to create such violence as a lightning flash. Naturally, if the separation between the positive and negative bodies is smaller than the separation between the thunder cloud and the earth, less voltage is necessary to produce a spark discharge. The truth of this principle is found in the motor car's spark plugs. Furthermore, if the atmosphere is so rarefied that the electrons and ions can move at destructive speeds, a discharge can occur. It will be accompanied by light for the same reason as lightning.

The electrons moving at such rapidity disturb the particles of air or gas through which they speed, thereby causing ionization or breaking up of the molecules. The effect produces a flash of light. In the case of lightning water vapor heated to the explosive point forces the air out from the path of the spark discharge and when the air rushes back into the vacuum pocket there is a terrific roar. This does not occur in a neon tube because no water vapor is present. The neon gas is ionized only to the extent that it produces a luminous effect without noise.

Fortunately for radio a rarefied atmosphere can be brought about in a sealed glass bulb and two electrodes provide the opportunity for the spark discharge or miniature lightning flash to occur. A bulb of this type as developed in the Bell Laboratories contains a rarefied atmosphere of neon—a chemically inert gas. A voltage is applied to the electrodes and a glow discharge results. It continues as long as the voltage is applied. That is why the neon tube is often called a glow discharge lamp. The brilliancy, after sufficient voltage is applied to make it glow, depends directly upon increase in the voltage. The lamp is kept "alive" by the local source of voltage and when additional impulses transmitted by the distant electric eye reach it, the brilliancy corresponds to the addition in voltage. The glow is seen on the plate of the neon lamp.

At any instant, only that part of the plate which is exposed by the moving aperture of the scanning disk can flash light to the eye. Despite the fact that the luminosity at any moment is uniform throughout the lamp, if from instant to instant it varies in the same way as does the light and shade of the successive squares of the picture of a televised person, then the spectator looking at the neon lamp through the hole in the scanning disk sees a picture.

EXPOSURE IS INSTANTANEOUS.—Mills asked his audience to imagine that the neon bulb is actuated by a single photoelectric cell placed as an eye before a distant scene. Suppose this cell is shielded so that it receives light from only one detail of the scene at a time. In that case the neon tube corresponds in brilliancy to one detail or tiny square of the scene. Then if the cell is successively exposed to the light from all the small areas of the scene, sweeping it from left to right, row upon row, in the same order and at the same speed as the aperture of the scanning disk whirls in front of the neon tube, the observer looking at the lamp sees the scene in entirety.

Obviously, similar mechanical arrangements are essential at the sending and receiving stations. The photoelectric cell at the sending end should be shielded so that it is exposed to only one spot of the scene at a time. A shield at the receiving terminal exposes to the eye only one spot at a time on the neon lamp's plate.

In television a photoelectric cell is exposed to each detail of a scene for only about one fifty-thousandth of a second; therefore, intense illumination is needed. This necessity is well illustrated by photography. The so-called instantaneous exposure needs bright illumination, strong sunlight or a flashlight. The scene must be flooded with light. Less illumination is needed for a time exposure. In television, however, too strong a light is inconvenient for the actors even though the photoelectric cells require it. The solution of this difficulty is based on the physiological phenomenon that it is not so much the instantaneous intensity of light that bothers as it is the average intensity to which a person is exposed. The quick flash of intense light produces little inconvenience, because of the sluggishness of that physiological process of sight exemplified in the persistence of vision.

QUICK RELAY TO THE BRAIN.—Once the television system has done its work and the image appears on the screen it is relayed to the brain by the eyes which comprise a lens system, a sensitive retina and an optic nerve. An image of the object being viewed is formed on the retina by the lens. All the light that enters the eye from any point is brought to a focus at a point upon the retina. Just as in the camera, where a lens forms an image on the film or plate, the intensity of the light, which the lens concentrates upon each tiny portion of the retina, depends upon the intensity of the illumination of the corresponding portion of the object or scene.

The retina, however, is not the smooth surface that it appears to be. Neither is the photographic plate or film, which consists of fine grains of a light-sensitive chemical.

ACTION OF THE RETINA.—"The retina consists of a surprising number, many millions, of fine rods and cones, of which the forward ends form the surface for the image," said Mills, "and the rear ends make the connections with an equally amazing number of nerve fibers. And these in a bundle, known as the optic nerve, pass from the eyeball to the brain.

"Through the almost innumerable channels of the cable-like optic nerve the brain receives simultaneously, but separately, all the reports of the illuminations to which each of the sensitive elements of the retina is exposed. Each transmits a stimulus proportional to the light falling upon it and varying therewith. Each is a light-sensitive element which sends out along its associated transmission line a current which produces in the brain a corresponding effect. Because of the many channels and the complexity of the brain it is possible for all the light-sensitive elements to transmit simultaneously and for the brain to perceive as a result a scene or picture."

The miraculous faculty of the human eye is shown in what man would have to do to pattern a system of television after the eye. It would consist of a myriad of small but wonderfully sensitive photoelectric cells upon which a huge lens would form an image of the scene. And each cell through a separate circuit would transmit to a small but efficient neon tube. Thousands of cells and tubes would be required, even if they could be made sensitive enough to operate and small enough to fit closely together in simulation of the finely compact cones and fibers of the retina. Moreover, a huge cable with thousands of wires would be required for connection between cells and tubes.

How the Eye Is Fooled.—"Insensitive and crude as are the photoelectric cell and neon tube in comparison to the corresponding elements of the retina and brain, their combination has one superiority," said Mills. "Their action is essentially instantaneous while the physiological elements have a tardy action. Flash a light for an instant before a photoelectric cell and its current makes an instantaneous surge. Repeat the flash twenty times a second and the same number of times the current from the cell will rise sharply to a peak and as abruptly fall to zero. On the other hand, repeat the experiment for a physiological eye and its brain will perceive only a continuous light. The effect of the first flash persists and, provided the next follows soon enough, no discontinuity can be perceived.

"It is this phenomenon, known as persistence of vision, that underlies man's ability to receive an illusion from motion pictures. Separate pictures are flashed on the screen at the rate of fifteen or more a second. Between times the screen is dark. But what man perceives is a screen continuously illuminated by a picture, the scenes of which change in an apparently natural manner like those seen directly. The two dissimilar phenomena, namely, persistence of vision for the physiological elements and instantaneous operation for the physical elements of cell and tube, are utilized in television."

SCANNING IS AN OLD ART.—The term "scanning" is not unfamiliar. The human eye affords a perfect example of scanning. Hollis Baird once explained television scanning in a way that makes it easy to understand in relation to the eye.

He said that without thinking analytically about it, when a person looks at a picture or a scene he sees it all at once, but the fact is that only a tiny spot is seen. What happens is that the flexible, efficient eyes rapidly travel across and up and down a scene, registering the various points so rapidly that a complete picture is observed. It is easy to test this.

Hold your hand out straight in front of you and look at the thumb nail. Now without shifting the eyes in the slightest note what else can be seen clearly, not just suggested, but vividly. You will find that the area comprising the end of the thumb is about all that is sharp. Now open the hand and decide that you want to see all of it. As you do, notice carefully what the eyes are doing. They are swinging back and forth in various cross directions until they have covered every bit of the hand. You have a definite picture of what the hand looks like, yet it was obtained piecemeal.

Taking something more concrete, more nearly like what a television camera must pick up, consider a motion picture. As the action goes on you seem to see what is happening on the whole screen but if you pick out a single spot on the screen and look at it without moving the eyes, as you did when looking at the thumb nail, you will discover that you are actually seeing but a small part of the picture clearly, the rest being in sort of out-of-focus relation to the main spot of vision. The human eye, however, moves so quickly that it takes the whole picture in a series of rapid glances and the memory retains these pictures, each piece in its proper place. The effect is a complete picture.

In television the same thing takes place. The television camera rapidly scans the scene which in turn is reproduced in the same order by the television receivers. This scanning is more rapid than the human eye, because the scanning spot cannot pick up as much detail as the eye will register correctly at one instant and so must travel faster to get in all the points.

The eye needs no definite routine to follow in scanning a scene. It may move across the top, then down to the bottom and across there, then up at an angle from the lower left to the upper right corner. In television, however, as in anything mechanical or electrical, an accurate pattern must be followed to be repeated in rapid succession so that the same pattern may be followed and reproduced at the receiving end. While television is a far cry from any human parallel, it actually follows the eye more accurately in its action than does a camera which takes in the complete picture at once.

The wonder of the human eye and ear stands out in bold relief when compared with man's radio-television system, which is bulky, cumbersome and relatively crude despite its magic performance. One has only to compare the delicacy, refinement and size of the mechanism of the eye and ear to realize that the most miraculous part of the entire television system extends from the eyeball and the eardrum to the brain. If man could pattern television after the eye and ear, radio instruments would be compact, tiny devices, no larger than an earphone, no heavier than a pair of glasses.

CHAPTER THREE

AN ELECTRIC HEART THROBS IN SPACE

Glass bulbs pumped free of air, with their miracles intensified because of the mysterious vacuum within their thin walls, are the heart of the television system. They flood the wireless circuits with the life-blood electricity that starts electromagnetic waves on their way to fling faces and scenes across mountains, over seas and through the very walls of the home, whether it be a hut in the mountains of Tennessee or an apartment on Manhattan Island. The electron tube does the trick. Man is constantly learning more about it.

Some day there may be in general use for reception a "cold," filamentless device that functions by chemical action or by a gas under pressure in a narrow barrel-like tube that looks more like a fountain pen than an electric lamp. A small battery might supply the current instead of the electric lighting mains. And there may be mercury vapor tubes.

IT'S ALL IN THE VACUUM.—Television images have in their veins and arteries specks of electricity valved by the central organ of the system—the vacuum tube. Through it a new realm has been discovered in the kingdom of science. It is called electronics. It seems to have no frontiers and so it fascinates all the more. It is all-powerful, invisible and quick in the performance of its wonders.

Electronics is a scientific force that grips the earth and plays a vital rôle in the everyday life of man. Electrons turn the wheels of industry and rush waves out into the emptiness of space far beyond the orbit of the moon, seeking new worlds to conquer. The domain of the electron extends as far as radio's pathway runs, and no one knows where it ends. It encircles the globe from pole to pole and girdles the sphere at every latitude. Radio echoes that flash back from stellar space reveal that this science is unfathomed. There is no yardstick to measure the electron's possibilities. They seem to be endless.

Radio performs its magic as an invisible force until it pops up on the television screen. Messages of joy and sadness, images portraying comedy and tragedy speed above the housetops unheard and unseen until man beckons, and with the proper instruments at his command, bids them welcome. The miracle of changing the invisible waves into sounds to be heard and scenes to behold, is all done inside a glass bulb, devoid even of air!

The sophisticated inhabitants of this modern world are asking the engineer, "Wherefore and whither are we going?" and the answer has a touch of mystery, "It's the vacuum— it's all in the vacuum."

A Mysterious Something.—In the vacuum! But can there be anything whatever in a vacuum—defined as empty space, devoid even of air? Emphatically, yes! It was Dr. Willis R. Whitney, director of the research laboratory of the General Electric Company, who uttered the paradoxical statement, "The vacuum—there's something in it!"

It is this bulb with "something in it" that helps science to harness the power of nature and make seeing by radio stranger than fiction. The unobtrusive tubes, each enclosing one of these areas of "nothing-something," made it possible to reproduce the front page of a San Francisco newspaper at Schenectady, 2,500 miles across the continent, three hours after it dropped from the press. A new automatic recorder developed by Charles J. Young accomplished this, and was a step toward fulfillment of a desire expressed by his father, Owen D. Young, who once remarked that he hoped to see the front page of the London *Times* flashed with a zip into

AN ELECTRIC HEART THROBS IN SPACE

New York. The words are printed on a roll of paper, which automatically moves through the machine at the rate of about one-half inch a minute.

Vacuum tubes glowing softly in snow-covered huts at Little America in Antarctica kept Byrd and his companions in constant communication with New York, while they were down there in 1929. Electromagnetic vibrations stirred up in the south polar regions raced across the world, over oceans, jungles and continents, to find slender targets of wire hidden amid skyscrapers in the Times Square district. They never missed their mark! It was the vacuum tube that enabled members of the expedition, when they reached New Zealand on the way home, to talk with friends in New York, while all-America eavesdropped on their conversation.

This is what the Rochester *Times Union* said about it:

> The experiment was of unprecedented size. It linked two voices in conversation over a distance of more than 10,000 miles. The arrangements for this tremendous hook-up were described as seemingly simple as a local telephone call. . . . The experiment, while not wholly successful because of static in New Zealand, is yet astounding in its scope. When the mind considers the wide advance the test represents over all previous records of the kind, it is plunged into a realm of fanciful speculations. Why, it seems within the bounds of reason that some day we may have a machine which will shoot its strength out into the vast field of ether and bring back the thundering voice of Cæsar, the doleful singing of Dante, the ring of spears and swords before the gates of Troy—and even the dying groans of the giant that Jack killed!

An All-pervading Force.—Electronics is closely allied with radio and television. The vacuum tube would stop beating and the vastness of space would cease to pulsate with human thoughts and emotions, if there were no electrons. This Aladdin lamp would be no more potent than an empty milk bottle or a burned out incandescent bulb, if some strange force suddenly destroyed electrons. They are all-powerful, all-pervading, yet so small that they defy the most sensitive microscope to single out one of them and watch its action. So tiny are these electric specks that if a drop of water—which contains millions of them, because of the hydrogen and oxygen within it—were magnified to the size of the earth, each electron magnified in proportion would be about as large as a grain of sand. The electron is approximately 1,700 times smaller than the atom.

When the big tubes in a television or broadcast transmitter are cooled off and at rest, the electrons, too, are "slumbering" in the filament. But as soon as the current is turned on the filament glows and the electrons leap from their reveries to perform useful work for mankind. Scientists point out, however, that electrons merely rest, that is, comparatively speaking. They never sleep. They are always moving back and forth at high speeds in the materials they occupy. But in the radio tube they leap with great velocity when the filament is fed with current to heat it. They rush away at the rate of approximately 50,000 miles in a second.

They instill life into the radio circuit and into the arteries of television. Power surges into the aerial wires. Space is made to vibrate with music, voices and images that are crisscrossed through the air. One little radio bulb has been known to hurl a message around the globe, shaking the great expanse of the earthly envelope as easily as a bowl of jelly can be set in motion by the tap of a finger.

But what is this invisible medium that shakes or vibrates when the electronic tube oscillates to send forth a message or image? Some call it the ether, an unseen, odorless, tasteless substance, believed to occupy all space. Others doubt that such a medium as the ether exists. Nevertheless, scientists agree that there is some marvelous force that lurks in the universe to complete the alliance of the electron tube with radio and television.

AN ELECTRIC HEART THROBS IN SPACE

EINSTEIN'S IDEA.—Dr. Albert Einstein has discarded the theory of the ether. He derides the radio's ethereal medium as fiction, calling it merely a makeshift fabricated to explain something for which scientists have not had the correct explanation. In an address at Nottingham University, he said that he believes radio's medium is an electromagnetic phenomenon. So did Charles Proteus Steinmetz.

"It now appears that space will have to be regarded as a primary thing, with matter only derived from it, so to speak, as a secondary result," said Einstein. "We have always regarded matter as a primary thing and space as a secondary result. Space now is turning around and eating up matter. Space is now having its revenge."

Teachers and technicians try to convey the idea of radio's medium by comparing it to a pond of water. When the electron tube gets into action in the broadcasting circuit an electric current surges out into the aerial wire to cause invisible waves to vibrate in much the same way that a stone cast into the pond starts a series of ripples or waves. That simple analogy helps the layman to comprehend how radio programs and television faces reach him.

AMONG THE PIONEERS.—Electromagnetic waves have existed in some form or other since man first roamed the earth. Light waves are called electromagnetic. Different colors of light are waves of different lengths. Scientists, or men of a magic turn of mind, back in 600 B.C. observed that by rubbing a piece of amber and some wool together, little particles of straw jumped to the amber. The tiny straws leaped up in much the same way that iron filings or a needle are attracted to a magnet. Centuries passed before queer-looking machines were devised that would produce electric shocks Then the Leyden jar was invented in which to store electricity. Benjamin Franklin sent up a kite during a lightning storm and showed that electricity generated by man-made machines was of the same nature as atmospheric electricity.

It was Michael Faraday who suggested that possibly waves of light might be an electromagnetic phenomenon. He conducted hundreds of experiments over a period of years in an effort to discover a relationship between electricity and light. He found that polarized light could be affected by a magnetic field.

James Clerk Maxwell, in 1865, took Faraday's mathematical calculations and proved them to be correct. He declared to a doubting world that electromagnetic waves could be produced and made to travel through space at the speed of starlight. Maxwell did this mathematically. Hertz confirmed the theory by creating and detecting the waves.

THE BURIAL OF A THEORY.—The New York *Times*, on October 5, 1931, in an editorial gives a good account of Maxwell's contribution to science and the reasons for his theory of the ether:

> In honoring Clerk Maxwell on the hundredth anniversary of his birth, British science both digs a grave and erects a monument. The grave receives the remains of his theory of a luminiferous ether; the monument is to his mathematical genius, which ranks with that of Einstein.
>
> It was a necessary creation—this ether of Maxwell's. Like Faraday and others before him, Maxwell could not believe in "action at a distance." To see a star the eye must touch it in a sense. To attract a needle a magnet must be "connected" with it. Maxwell invented an ether that satisfied the conditions. It was a vastly different ether from any that had been postulated before. Not only did it transmit light, electricity, magnetism, but revealed them as different manifestations of the same primal, radiant energy. Just as Newton's laws of gravitation unified the heavens, so this new ether unified matter and energy. It explained everything but gravitation. Lord Kelvin could write of it in 1899:
>
> "The ether is the only substance we are confident of in

AN ELECTRIC HEART THROBS IN SPACE

dynamics. One thing we are sure of, and that is the reality and substantiality of the luminiferous ether."

Yet even then the ether had been molded to fit new discoveries. More tenuous than any gas, it filled the spaces between atoms and stars. It was as viscous as wax. It was a jelly capable of transmitting vibrations. It was subjected to strains that would snap steel like matchwood. It was so dense that a quantity no bigger than a pinhead would sink through iron as a stone sinks in water. It was full of twists, pulls and pushes. It formed vortices that we recognized as matter. In a word, it was the supreme paradox of Victorian science and yet a triumph of the scientific imagination.

That ether is gone. Its properties have been acquired by space—not Euclidian emptiness, but an emptiness strangely endowed. Gravitation falls neatly into line as a geometrical attribute of space and is satisfactorily accounted for. The universe is no longer a machine, but a problem in higher geometry. Maxwell's fate is much like Newton's. A slight modification of the laws of gravitation has given us a new universe, which is really simpler than the old, though it may still be incomprehensible to most of us. The mere transference of the ether's properties to non-Euclidian space has carried the unification of energies further than Maxwell thought possible. Reality acquires a new meaning in which he would have rejoiced. Were he alive he would probably concede that his ether was no more real than the "average man" of the statisticians or the equator of the geographers—that it was a necessary and convenient fiction without which the science of his day was helpless.

WHY THE ETHER?—Looking back to the days of Marconi's early triumphs, Einstein points out that at that time the only real things were bodies, space and time. Those were the constructive elements from the physical point of view. Had not Faraday introduced the idea of an electric or magnetic field, such as surrounds an ordinary magnet? Scientists, therefore, were called upon to introduce a new body called the ether to represent a physical state. This, theoretically, allowed the electromagnetic phenomena to occur in space.

"Looking back," said Einstein, "now we must ask why ether as such was introduced? Why was it not called 'state of ether' or 'state of space'? The reason was that they had not realized the connection or lack of connection between geometry and space. Therefore, they felt constrained to add to space a variable brother, as it were, which could be a carrier for all electromagnetic phenomena."

THE STEINMETZ DECLARATION.—It was not so long after broadcasting started in America and everyone was discussing the wonders of the ether, that the electrical wizard Steinmetz upset popular belief by his famous sentence, "There are no ether waves." He emphasized the fact that radio and light waves are merely properties of an alternating electromagnetic field of force which extends throughout space. Scientists, he contended, need no idea of the ether. They can think better in the terms of electromagnetic waves. And it was for this reason that the distinguished Steinmetz heralded the Einstein theory of relativity as "the greatest contribution to science of the last fifteen years."

Steinmetz, like Einstein, declared that the conception of the ether is one of those hypotheses created in an attempt to explain some scientific difficulty. He asserted that the more study is applied to the ether theory the more unreasonable and untenable it becomes. He contended that it was merely conservatism or lack of courage which kept science from abandoning the ethereal hypothesis. Steinmetz further pointed out that belief in the ether is in contradiction to the Einstein theory of relativity, because this theory holds that there is no absolute position or motion, but that all positions and motions are relative and equivalent. Thus, if science agrees that the theory of relativity is correct the ether theory must be cast aside.

AN ELECTRIC HEART THROBS IN SPACE

A MAGNETIC RESERVOIR.—The space surrounding a magnet is a magnetic field. To produce a field of force requires energy, and the energy stored in space is called the field. This is supposed to be an accumulation of the forces of all the electrons in existence. In radio or television the transmitter with its electron tubes disturbs the energy which is stored in the great reservoir of space which listeners tap to hear music and to see pictures. The globe is also surrounded by a gravitational field. When a ball is thrown skyward it falls back because it does not have sufficient force behind it to overcome the power of gravity which acts upon it.

If a coil of insulated wire is wound around a piece of soft iron and a direct current is sent through the coil it becomes an electromagnet. The space around the coil is the magnetic field. When the current is increased the magnetic field increases. When the current is decreased the breadth of the field is reduced. If the current is reversed the field is reversed. When an alternating current is sent through the coil the magnetic field alternates. The field becomes a periodic phenomenon or a wave, described by Steinmetz as "an alternating magnetic field-wave."

"The space surrounding a wire," said Steinmetz, "that carries an electric current is an electromagnetic field, that is, a combination of a magnetic field and an electrostatic field. If the current and voltage alternate, the electromagnetic field alternates; that is, it is a periodic field or an electromagnetic wave."

So today, the modern broadcast listener or television spectator who wants to forget the ether can visualize the aerial wire at the transmitter setting up electromagnetic waves in a field of electric force, which now, the theorists contend, fills all space and, therefore, every receiving antenna is within the field. The broadcast or television transmitter jars the hypothetical medium, causing it to vibrate. The greater the power of the transmitter the greater will be the vibration and the farther it will carry. The receiving set is designed to detect the vibrations, and accordingly intelligence and images are broadcast from one part of the world to another. Such is the power of electronics and of an "empty" glass bulb.

A SCIENTIFIC BURGOMASTER.—The pages of history reveal that Otto von Guericke, burgomaster of Magdeburg, was a pioneer in electrical science. His accomplishments included the invention of an air-pump with which he obtained a partial vacuum—not a high vacuum, such as is common today, but still one in which the air content was thin.

One day in 1654 he called by appointment on Emperor Ferdinand III, accompanied by two teams of eight horses each, with their drivers and various queer paraphernalia. He showed the Emperor two copper bowls which, when placed together, formed a hollow sphere. Between them von Guericke inserted a ring of leather soaked in wax and oil, making an air-tight joint, but there was no mechanical connection whatever. With his air-pump he drew off a great deal of the air from the sphere through a hole which was closed by a tap.

The teams of horses were then brought up, one being hitched to each of the copper bowls or hemispheres. At the signal to go the sixteen horses pulled and strained, but their utmost exertions could not drag the hemispheres apart. The Emperor, amazed, found it impossible to believe that the bowls were locked together merely by the difference in air pressure between the atmospheric density outside and the partial vacuum within. This was the vacuum doing tricks.

EDISON ON THE SCENE.—The vacuum at work universally did not come until two centuries later, and Thomas Alva Edison was the scientific "magician" of this later performance. By that time men knew more about electricity; and there is a close working relation between electricity and the vacuum. Edison placed a carbon filament within a vacuum, and then connected the filament to an electric circuit. The resistance of the filament to the passage of the electric current made it glow with incandescent light, while the vacuum prevented it from burning up—and lo! the incandescent electric lamp was born, essentially a vacuum device.

Edison, as fate would have it, did more than construct a practical electric lamp depending on a vacuum. He was the first to observe a peculiar electric current originating with the hot filament inside the vacuum. Today it is known as "the Edison effect." He placed a metallic plate inside the lamp near the filament. Then he noticed, when the current was turned on, the filament became hot and the needle of a galvanometer or current indicator was deflected, despite the fact that there was no connection between the filament and the plate to complete the circuit. The electron stream was completing it. The commercial possibilities of electric lamps seemed more practical to the Wizard of Menlo Park, and he turned his attention to that field of research, leaving "the Edison effect" as a clue for others.

The present electrical age, so-called, is unfolding in astounding fashion, remarked an engineer at "the House of Magic." It has come to stay and may in time reveal successive distinctive epochs, like the geological eras in the age of the earth. The world has already passed through the magnetic-electrical epoch. Now it is entering upon the vacuum-electrical. Possibly this will be followed by the atomic-electrical, and that in turn by the cosmic-electrical, in which tremendous undiscovered forces in outer space will become servants of man. In that epoch a literal tour of the solar system may be achieved, and the world will gaily dispatch its interplanetary Lindbergh—a goodwill ambassador to the stars! Fantastic? Ah, but truth is stranger than fiction—and stranger than ever as the years pass.

PART II

THE MARCH OF TELEVISION BEGINS

TELEVISION

(Editorial in The New York *Times*, December 17, 1926)

> For he looketh to the ends of the earth,
> And seeth under the whole heaven.

This was one of the poetical statements used by Zophar by way of comforting Job in his many tribulations, in order to illustrate the omniscience and omnipotence of the Almighty. Job at last yielded, saying:

> I know that thou canst do all things
> And that no purpose can be restrained.

Once the scientist said in the mood of Job, "With God all things are possible," but these are things "too wonderful for me, which I knew not." Now he is unwilling to say that there is anything impossible with man. Speech at great distances was for ages never thought of as a possibility, and, even after communication by wire was achieved, not dreamed of as feasible without the assistance of wire. Sight at great distances has at last been made possible by telephotography, the carrying of images across thousands of miles.

Now comes in prophecy of actual achievement the almost instantaneous flight of images in motion across seas and continents, just as Lucretius, nearly two thousand years ago, explained their movement in his theory of the visibility of objects near and far: the air being filled, as he conceived, with millions of images, ever passing and crossing one another in every direction, some swifter, some slower, in infinite complexity, yet in no confusion, "very unsubstantial," yet "keeping their forms as they speed on their way to the senses." He went even further in describing these as being traversed by images of the mind, and these in turn by the majestic images of the gods. But the amazing thing is that images do now actually cross one another in every direction and in "infinite complexity" and yet keep their forms intact and become visible to the eyes thousands of miles away.

Television is an accomplished fact by means of radio photography, but it remains so to quicken the process of transmission as to make moving objects visible in life size on a screen at a distance. What is required, in the language of a related art, is a brush of light that will more swiftly bring these images into view. As Dr. E. F. W. Alexanderson of the General Electric Company explained to the American Institute of Electrical Engineers, it will be necessary to increase the operating speed from 40,000 to 300,000 picture units per second in order to get pleasing results. The "brushes of light" have been multiplied. Even so the "painting" cannot be done rapidly enough. And there seems to be a question whether mechanical power can be sufficiently swift to recover these images. But that in some way, if not in the mechanical acceleration of these brushes, the thing will be done cannot be doubted. What needs to be done being known, the way will be found. For that confidence, we have the support not only of past achievement but of the eager and never-satisfied effort of the human mind.

Science's search will continue till it can say as Job did at the end of the greatest interview in all literature between man and the Voice of the Whirlwind:

> I have heard of thee by the hearing of the ear:
> But now mine eye seeth thee.

Chapter Four

EXPERTS ANALYZE THE PROBLEM

Let us go into the darkened television studio of Alexanderson in the Mohawk Valley, or into Zworykin's scientific sanctum, where the big radio eyes look down on the visitors.

Step into the mystic laboratory of Ives, where in a darkened booth he invites his guests to glimpse through a peek hole to behold a bouquet of flowers in color, and the Stars and Stripes waving in all its glory with the red, white and blue as natural as if floating from a mast in the noonday sun.

Go with Jenkins and see his images dance on a screen. Listen to the fascinating story of John Baird, who sent the sound of a face across the Atlantic from a mysterious room in London, to be picked up in a dark cellar on the outskirts of New York.

Television is in the news!

Is Privacy Menaced?—When Marconi and other scientists first predicted that radio vision was in the process of evolution the layman feared that a simple all-seeing device with piercing eyesight was destined to strike a deadly blow at privacy.

New Yorkers visualized neighbors and even the residents of California looking through the walls of the apartments on Manhattan Island. They reasoned that if a spectograph could observe the action of electrons gyrating in metals at a speed of 90,000 miles a second it might be an easy task to build an electric eye capable of peering not only into the home but into the mind of man!

All these illusions are cast aside and fears put to rest

when it is explained that television requires an electric eye or radio camera in the home before the family cheer or troubles can be aired.

There is no better way to follow the spectacular march of television than to listen to the historic utterances of those who have nurtured the images from hazy, spiritual-looking things to clear-cut faces that live with a personality of their own. They are no longer flimsy, fading images but life-like characters with plenty of strength to climb up the ladder of science to aerial pinnacles from which they leap unencumbered to the homes of all the land. Each announcement of progress by the inventors, each lecture and demonstration of a step forward, when knit together as a running story discloses the romantic tale of television in its battle against the elements as scientists delve into the secrets of nature, chemistry and electricity. This inquisitiveness on the part of man enables him to learn how to build delicate instruments so that moving pictures are unfettered for a flight through space without surrendering their identity.

To sit down with these men in their laboratories, or to be with them in their leisure moments of recreation when they talk television, because they cannot dodge its magic spell, is to hear a running history of how a new scientific art evolves. They have had weird experiences.

They have seen radio "ghosts." They have watched their images travel to the antipodes and back in the fraction of a watch tick. They have seen faces pass through a skyscraper and come out the other side mangled and tangled beyond recognition, with an ear missing or with a side of the face gone, absorbed by the lattice steel structures that reach aloft like giant metallic fingers to pluck energy from the fleeting waves.

They have seen nature freckle a face by bombarding it with static in much the same way that a boy ruins the countenance of a snow man by throwing pebbles at it. They have watched their images being spoiled by nature in causing the waves to wax and wane as the invisible impulses encountered mountains, hills and valleys.

Then, too, the faces are often blurred, distorted and wavy like a picture printed from a film the emulsion of which was moved in streaks before it dried. And they have tenderly released the images from aerial masts never to see them again—not even a trace—because some cruel force in nature led them astray.

The drama of television unfolds in a most magnetic way as the inventors spin the historic yarn by their announcements of success, by their public lectures that reveal startling discoveries. Their weird observations disclose how nature has hidden and protected certain scientific facts throughout the ages, held in bondage until man was ready to seek and to harness them for a useful purpose. And it will be noted that in tackling scientific problems and in striving to overcome strange obstacles, man usually approaches from a complex angle. He conceives complicated devices.

In the end, however, a simple instrument generally solves the baffling problem and man smiles to see how really simple is the answer. Television today is less complicated than the experts thought it could be back in 1920.

Now let us follow the march of television step by step, in chronological order, and in as non-technical language as possible, because there is no better way to observe and to learn how the miracle is performed.

Marconi Expects a Visible-'phone—May 22, 1915

King Victor Emmanuel of Italy requests Guglielmo Marconi to return to his native land because of Italy's entrance into the World War. And so he sails on the steamship *St. Paul* of the American Line bound from New York to Liverpool, whence he will go across France to Rome.

Prior to sailing the inventor announces that engineers are

working on a wireless device by which a person can look through a solid wall. It is said to resemble a camera, which, when placed against a wall or floor, makes the wood, stone, bricks, concrete or metal transparent—in this respect resembling the X-ray. He says the instrument is not perfected, nevertheless, persons can be seen in the next room if they are close enough to the wall, but the image is blurred if they are a little distance away.

"And the visible-telephone—where persons talking can see each other—is coming successfully," said Marconi, "although I am not working on it."

The public is wondering what the wizardry of wireless will do next.

Up from the Graveyard of Ideas—June 3, 1925

In the evolution of sending pictures by wire and radio, a step that leads to television, there has been built quite a graveyard of ideas. Eighty years passed from the inception of transmitting pictures and facsimile dispatches by wire before commercial application was practical. This long-pull development was due to the fact that it is inherently more difficult to send a photograph than to transmit a telegraph message or the voice.

Captain Ranger, in a lecture before the Institute of Radio Engineers, called attention to the fact that Samuel F. B. Morse's contribution to communication was not alone, as most seem to think, the development of a telegraphic instrument, but largely the development of the telegraph code. Any number of telegraph devices had been constructed before Morse, but they did not have the economic practicability of an all-round system which would get words across to a distant point in a short period of time.

"How successful Morse was may be realized, when, today, it is an established fact that the Morse code, representing letters by dots and dashes, is still the most economical way

EXPERTS ANALYZE THE PROBLEM

of sending a given number of words from one point to another, in the shortest time, with the least power, over the greatest distance, and through maximum interference," said Ranger. "Of course, other means of sending words have been produced, typically, the telephone; but it requires a higher quality of wire service and perfection in apparatus to accomplish the high speeds attained when words are transmitted by voice.

"As soon as we understood the economic angle of the problem of sending photographs, we began to look for a picture shorthand. The whole problem was largely one of realizing what confronted us and what our real aim was. Then the answers began to come easily.

THE PICTURE IS CUT UP.—"Practically every system to date has been, and still is, on the basis of dividing the picture into small unit areas and to transmit their values one after the other. When we stop to think that the usual newspaper half-tone has at least sixty-five dots in a row for an inch, or more than 4,000 dots to a square inch, the magnitude of the job becomes apparent. The usual method of picture transmission has found its serious drawback in the number of pulses that have to be put through; and the precision with which they must be sent; and the time that it takes to send them."

Search for a shorthand method was started. The first effort in this direction consisted of variable dot-spacing. Obviously, if dots are placed on a piece of white paper and spaced widely, they give an impression of white. If they are placed close, black is approached. That is what was done in the first shorthand attempt, making each dot of generally the same size; although it worked out that the individual dots widely spaced were a little lighter than those grouped together. These dots by their grouping constituted the shades of the picture.

A TRANSOCEANIC TEST.—The first public transatlantic demonstration of the transmission and reception of pictures by radio, utilizing the Ranger method, took place in November, 1924. The photoradiogram transmitter was located in London. The signals from this apparatus were put on the 220-mile land line to Carnarvon, Wales, at which point they actuated the control relays of the high power radio transmitter there. The signals from Carnarvon were picked up at Riverhead, Long Island, amplified, and sent by wire to the New York office of the Radio Corporation of America as audio frequency dots and dashes. The tone signals were again amplified at New York, then rectified and applied to the photoradiogram received.

JENKINS CALLS IT SIMPLE—SEPTEMBER 13, 1925

So definite is the progress being made in television that not so many years from now practically every household will have an attachment to its radio set, whereby the family will be able to see in the home events taking place at a distance. This will include the World Series baseball games, Presidential inaugurations and the Mardi Gras at New Orleans, according to C. Francis Jenkins.

This Washingtonian says that it does not seem strange to him that we shall presently plug into the loudspeaker jack of the radio receiving set a small box-like device which will project on a small white screen an action picture of some event taking place downtown or in some more distant city, a ceremonial, a national sports event, a spectacular scene in the news. He doesn't consider it mysterious, or even difficult. It only seems that way because it seems impossible, and it takes time to work out the details. It is the development, the refinement of each separate element, that is occupying his attention.

"Let's see whether or not I am warranted in assuming that it is a simple problem, whether there is really any mystery

JOHN LOGIE BAIRD
The Scotsman who sent an image across the Atlantic in 1928 and later televised the English Derby.

DR. HERBERT E. IVES
Electro-optical Research expert, the first man to fly the Stars and Stripes in color on a television screen.

VLADIMIR ZWORYKIN
The cathode-ray tube with the flat end covered with a fluorescent screen upon which images appear at the receiver, after being electrically scanned.

C. FRANCIS JENKINS
Washington inventor who began to study television in the '90s. He radioed a picture of President Harding from the national capital to Philadelphia in 1923.

PHILO T. FARNSWORTH
The Californian who used the cathode-ray tube to serve as the heart of his novel television receiver. He is an advocate of electrical scanning.

in the thing after all," said Jenkins. "Let's analyze the problem; take it to pieces and examine it in detail.

"These are the essentials. We want a picture of a remote scene. We want it repeated fast enough to produce the motion and we want it carried into our homes from the distant baseball park, let's say. That's the problem, and that is all there is to it, namely, a picture of a distant activity.

"If a man puts his head under the black cloth of an old-fashioned camera pointed at the baseball game he sees in miniature on the ground glass an exact reproduction of the game as played. It is carried by light from the baseball diamond to the ground glass screen. That is exactly what we want, only we want it in our homes. So light working alone won't do, because light goes only in straight lines, and obstructions cut it off; we must, therefore, have some sort of a carrier which can go around obstructions and through the walls of our houses. A copper wire will do, but a wire carries only to one place. So let's take radio! That carries everywhere.

A Boyhood Trick Recalled.—"Now we come to the consideration of the picture," continued Jenkins. "A picture is nothing but some black and white mixed up in a definite order. Pick up a modern photographic portrait, which, by the way, is the almost perfect example we have of the delicate blending of light and dark and half-tones. Examine it analytically and you will see what I mean. But how are we going to make radio, which has carried these lights and shadows from the ball park to our home, reproduce the ball game as the picture?

"That's easy!" exclaimed the inventor. "Don't you remember when we were little tykes mother entertained us by putting a penny under a piece of paper, and, by drawing straight lines across the paper, she made a picture of the Indian appear. Well, that's the very way we do it.

"So, in our homes we take a desk square of white blotting

paper and we move across it in successive lines an image of a small light source. If this little light spot moves across the screen swiftly the eyes see it as a line, like the circle of fire of our youth when we swung a lighted stick. Now, when these successive lines, one under another, are made so swiftly that the whole screen surface is covered in one-sixteenth of a second we have motion picture speed, and the entire screen is illuminated.

"If, then, the incoming radio current is put through our lamp, the strong signals will make the spot of light on the screen very bright. The weaker signals make it more dusky and when there are no signals the lamp goes out and the screen is no longer uniformly illuminated, but the light is dabbed over the screen. And because a picture is only a collection of these little dabs of light put around in different places on the screen, it will readily be seen that these radio light variations, when they follow a predetermined order, make up our picture of the ball game, just as the humps on the penny made up a picture of the Indian, although the pencil moved over the paper in straight lines.

"So that's the way we make radio pictures and radio movies in your home. The incoming radio signals turn the light up and down as it moves swiftly over the screen, and you 'see' the distant scene. Easy, isn't it? You can go out in the woodshed and build yourself one now. Of course, if you have only a fine laboratory and no woodshed where you can get off by yourself and think clearly you are out of luck. So, if you have a woodshed, go to it and good luck to you. If your woodshed is on a farm the probability of clear thinking is greatly enhanced."

Galloping After the Images—April 25, 1926

More than a dozen inventors teamed with a corps of expert assistants, many of them specialists in radio, electricity, chemistry and optics, have entered the race which will award

EXPERTS ANALYZE THE PROBLEM

the winners fame and possibly fortune in television. Alexanderson, a Norwegian by birth, but now an American citizen, represents the United States along with Zworykin, Jenkins, Ives, Farnsworth, Sanabria and Hollis Baird. Dr. Alexandre Dauvellier, Belin and Holweck carry the colors of France, while Denoys von Milhaly is in the contest for Austria. Baron Manfred von Ardenne, Karolus and the house of Zeiss Ikon are doing their bit for Germany. John Baird is in the race for the Union Jack.

Von Ardenne is developing the cathode ray method, and Dauvellier is an expert in cathode ray television. Incidentally, Boris Rosing of Russia is said to have originally proposed the use of cathode ray tubes in a television system which he patented, but that was so many years ago that the patents have expired, indicating that cathode ray television is no new art.

"My televisor is nothing like photoradio or telephotography," said John Baird. "The transmission of photographs or still pictures onto a plate is no longer a novelty. What the televisor does is to transmit to the human eye living and motion pictures at the instant of their occurrence. The problem has not only been that of converting light into electricity at the transmitter and reconverting radio waves into light at the receiver. The solution of that problem is nothing new. The big task has been synchronizing of the converting and reconverting processes and of speeding them up so as to give the eye the impression that it is seeing a whole picture instead of a succession of parts. Once these puzzles have been satisfactorily solved, we can broadcast motion pictures to any distance that wires or wireless cover. We can focus the lens of the transmitter just as a kodak is focused, so that the day will come when we can send not only the close-up of a face but a distant view of a battle in progress. It is all a matter of speed and proper synchronization of the instruments."

INTRODUCING "STOKIE BILL."—Baird interrupts his description to give a demonstration. He stands in a flood of light. A mop of curly, corn-colored hair tumbles over a wide brow and down the back of his neck over the collar to his rough tweed jacket. He closes a switch and a disk revolves at a whistling speed.

"Stokie Bill" lies on the window sill at his elbow. "Stokie Bill" is the head of a ventriloquist's dummy, and its garish likeness has been telegraphed ever since inventors began to develop telephotography. "Stokie" is a sort of mascot among inventors who work on the problem of picture transmission and broadcasting of images. Felix the Cat is assigned a similar rôle in the United States. These dummies perform on turntables and move about in front of the televisor's eyes for many hours under the glaring lights without the tiring effects that a human head experiences.

"There is only one thing that makes the problem of television an extremely difficult one," said Baird. "That is the speed of signaling which is necessary if we are to see an event at the moment at which it occurs. The transmitting and receiving mechanisms must not only be so sensitive in response to extremely dim light, but they must act instantaneously. Aside from the speed and synchronization, the problem is relatively simple.

"The general theory is to project a picture onto a light-sensitive cell in a piecemeal fashion. Each of the small areas into which the picture is divided causes the light-sensitive cell to send out an electrical current which is proportional to the amount of light in its 'area.' Thus the dim parts of the picture send out a weak current and the bright spots are represented by a stronger current. Then at the receiving station these currents control a source of light which is projected onto a screen in exact synchronism with the projection of the picture at the transmitter. The process is per-

EXPERTS ANALYZE THE PROBLEM 57

formed so rapidly that, due to the eyes' retention of the images, the whole picture appears simultaneously.

"The light-sensitive cell is nothing novel among inventors. I use only one cell at the transmitting end and I break up the picture into 'areas' by means of lenses in the whirling disk. The lenses in the disk focus the 'areas' of the picture, one by one, onto the light cell, and when the disk has been whirled once every 'area' of the picture or face has been focused consecutively onto the cell.

SECONDS ARE PRECIOUS.—"It is simple enough merely to transmit the 'areas,' but you must remember that we have to send them ultimately to the human eye. For instance, let us say that we take as much as half a second to broadcast a picture of a face. By the time the light-sensitive cell is transmitting the light values of the chin the eyes which are watching the screen at the receiving end will have lost the light values of the hair, and the result will be that, although our transmitting method in itself may be perfect, the eyes at the receiver will retain no image at all.

"We must be able to broadcast all the 'areas' of the face within a tenth of a second if the eyes at the receiving station are to retain the image of the face as a whole. To do this has been one of television's baffling problems. Once we have succeeded in overcoming that obstacle, we can transmit moving pictures as easily as the cinema does. Having given the eyes at the receiving end one complete picture in a tenth of a second, we can give it another complete picture in the next tenth of a second by merely keeping the disk whirling at the right speed at the transmitter. That is the ordinary cinema principle. It consists of an extremely rapid succession of still pictures.

"Practical television, therefore, boils down to the rapid transmission of light dots and a synchronizing mechanism. Suppose we want to broadcast the picture of an object in motion, say, two inches square. We must transmit at least ten complete pictures of it every second, and by the most conservative estimate this requires the transmission of about 25,000 light dots a second," explained Baird. "That is what my mechanism does. For the light at the receiving end I use a glow lamp, and for my synchronizing mechanism I move the spot of light across the screen by means of a slot and a rotating spiral."

THE INVENTOR'S WILL-O'-THE-WISP—DECEMBER 15, 1926

Television is called an inventor's will-o'-the-wisp. A light-brush is needed that will empower a beam of light to brush or paint about 300,000 image units per second on a screen. Such speed is inconceivable with electro-mechanical apparatus. The moving parts would fly asunder. Even if mirrors could be rocked or rotated thousands of times a second, there would not be sufficient light to illuminate a large screen effectively with life-size images.

The inventors are aware that the television screen to win public approval for practical home use must be larger than a handkerchief. They explain that the same holds true in television as in painting the side of a house—the larger the surface to be covered with a given amount of paint the thinner must be the coat; the larger the television screen to be painted by a light beam of given intensity, the dimmer will be the illumination.

How ALEXANDERSON REASONS.—Alexanderson, in a lecture at a meeting of the St. Louis section of the American Institute of Electrical Engineers, announces that he has solved the problem. With the ingenuity and simplicity of a great inventor, he reasons: "If one beam of light cannot brush a light-picture fast enough I will use several beams and divide the work among them. And several beams will give me several times as much light as one beam, so that I can brush images which will be both large and brilliant."

To do this he has a new television projector. It utilizes a

revolving drum carrying twenty-four mirrors which throw a cluster of light beams on the screen. As the drum revolves once a single spot of light passes over the screen twenty-four times, line by line. Seven spots give him a total of roughly 170 light strokes in one revolution of the drum. When the machine is idle, but the lights turned on, the seven bright spots appear as a cluster on the screen. As the drum whirls the spots move quickly. They gyrate and blend as they trace seven lines of light simultaneously, then another seven, and another seven until the entire screen is flooded in light. Thus seven crude pictures are simultaneously light-brushed on the screen with such rapidity that the eye has no time to follow the interlacing process and, therefore, obligingly combines them into a single good image.

"Our work has already proved that the expectation of television is not unreasonable," Alexanderson declared at this St. Louis meeting, "and it may be accomplished with means that are within our possession at the present time. How long it will take us to attain practical television I do not venture to say. It is easy enough to design a television system with something like 40,000 picture units per second but the images so obtained are too crude. They have no practical value. Our work in radio photography has shown us that an operating speed of 300,000 picture units per second is necessary to give pleasing results. This speeding up of the process is unfortunately one of those cases where the difficulties increase by the square of the speed."

Half-tone effects are produced by dividing the picture into five or more separate shades, such as white, light gray, medium gray, dark gray and black. The transmitting and receiving machines analyze and reassemble these shades automatically. The engineers have found various methods for translating light intensities into radio signals. One method is to use five wave lengths, one for each shade. However, in this Alexanderson process a single wave length is utilized.

MACHINE SELECTS THE SHADES.—The transmitting machine is made in such a way that it automatically at every moment selects the shade that comes nearest to one of five shades, and sends out a telegraphic signal which selects the corresponding shade in the receiving machine. This sounds more complicated than it really is, because the telegraphic code by which the different shades are selected depends upon the synchronization of the two machines, which is necessary under all circumstances. Thus, black in the picture is produced by exposure of the sensitive paper to the recording light spot during four successive revolutions, whereas light gray is produced by a single exposure during one of the four revolutions and no exposure for the three succeeding revolutions. The overlapping exposure is progressive and the whole works as a continuous process.

The television projector consists of a source of light, a lens and a drum carrying a number of mirrors. When the drum is stationary a spot of light is focused on the screen. The spot of light is the brush that paints the picture. When the drum revolves the spot of light passes across the screen. Then as a new mirror, which is set at a slightly different angle, comes into line the light spot passes over the screen again on a track adjacent to the first, and so on until the entire screen is covered with illumination. If a light-picture of fair quality is to be painted, at least 10,000 strokes of the brush are necessary. This may mean that the spot of light should pass over the screen in 100 parallel paths, and that it should be capable of making 100 separate impressions of light and darkness in each path. If this process of painting the picture over and over again sixteen times in a second is now repeated, it means that 160,000 independent strokes of the brush of light in one second are required. To work at such a speed seems at first inconceivable; moreover, a good picture requires really a scanning process with more than 100 lines. This brings the speed re-

EXPERTS ANALYZE THE PROBLEM

quirements up to something like 300,000 picture units a second.

Besides having the theoretical possibility of employing waves capable of high speed signaling, there must be a light of such brilliancy that it will illuminate the screen effectively, although it stays in one spot only one three-hundredth of a second. This has been one of the serious difficulties because even if the most brilliant arc light is employed, and no matter how the optical system is designed, it does not give sufficient brilliancy to illuminate a large screen with a single spot of light. Therefore, Alexanderson has built a new television projector in order to study the problem and to demonstrate the practicability of a new system which promises to give a solution to the difficulty.

The result of this study is, briefly, that, if he employs seven spots of light instead of one, he gets forty-nine times as much useful illumination. Offhand, it is not so easy to see why there is a gain in light by the square of the number of light spots used, but this can be explained by reference to the model. The drum has twenty-four mirrors, and in one revolution of the drum one light spot passes over the screen twenty-four times, and when seven sources of light and seven light spots are used there is a total of 170 light spot passages across the screen during one revolution of the drum.

Tests have been made with this television projector to demonstrate the method of scanning the screen with the seven light beams working in parallel simultaneously. The seven spots of light may be seen on the screen as a cluster. When the drum is revolved these light spots trace seven lines on the screen simultaneously, and then pass over another adjacent track of seven lines until the whole screen is covered.

A complete television system requires an independent control of the seven light spots. For this purpose seven photoelectric cells are located in a cluster at the transmitting machine and they control a multiplex radio system with seven channels. Seven television carrier waves may thus be spaced 100 kilocycles apart, and a complete television wave band should be 700 kilocycles wide. Such a radio channel might occupy the waves between 20 and 21 meters. If such use of this wave band will enable Americans to see across the ocean, Alexanderson believes all will agree that this space in the ether is assigned for a good and worthy purpose.

PREDICTING THE FUTURE.—"No one can accurately predict just what the future of radio television will be," said Alexanderson. "The inventors who gave us the moving pictures certainly never foresaw the time when film plays would be produced at fabulous cost and 10,000,000 people a day would pay from ten cents to two dollars each for the privilege of seeing Douglas Fairbanks and Charlie Chaplin on the screen.

"With the telephone it was the same. Many thought that the telegraph would be completely displaced, but the telegraph is as necessary as ever and the telephone now occupies a field of its own. Edison realized that the phonograph could preserve the voices of great singers and the interpretations of noted violinists and pianists for future generations, but in the early 'eighties no one dreamed that records by Metropolitan opera stars or ragtime and jazz by dance orchestras would be sold by the million.

"For these reasons I hesitate to become too televisionary. The apparatus which we hope eventually to build will be just as serviceable in transmitting motion pictures to the home by radio as in exhibiting news events directly. It seems certain that just as we have succeeded in combining sound records with motion pictures so that we can hear the words that photographed lips form, so television will be combined with broadcast music. Radio reception, as we know it today, is blind; television is deaf. Combine the two and we appeal to two senses at once, just as we do in any theater.

"Radio has already enriched the lives of thousands of

lonely farmers with music that was once heard only in the large cities. Ultimately it will be possible to receive in the village moving-picture theater a performance of *Hamlet* by John Barrymore or of the latest musical comedy that has captured Broadway's fancy. The curtain will go up and down just as it does on the stage in New York; the stage will be disclosed with all its scenery, but in black and white. Actors will be seen and heard in Wyoming as distinctly as in the theater itself.

"That the more important events will be picked up by wire, sent to the broadcasting station, and then radiated to television receivers within a radius of two hundred miles or more is a foregone conclusion," said Alexanderson. "Political conventions, state functions, the welcome of a queen to these shores, championship tennis matches and baseball and football games—all these will undoubtedly be flashed into millions of homes.

LOOKING ACROSS THE OCEAN.—"Seeing across the Atlantic Ocean will be no more difficult than hearing in New York a concert played in London. A vision will be sent across the water on several powerful waves and reradiated here on other waves used by our local stations. The practice is common enough now in long-distance transmission and reception of music and speech. If television is practical within a hundred miles of an American broadcasting station it is also practical in a transatlantic sense.

"The fundamental principles of radio communication and wire communication are the same," explained the inventor. "Visions can be sent over wires to specific destinations as well as through empty space. When television becomes practical we shall see the man we have called up on the telephone if it pays to see him. How important it is to gaze on him as we talk to him must depend on circumstances. It might be important to exhibit a murderer caught in San Francisco to the police in New York for identification without waiting for New York to send photographs or fingerprints.

"No one believed in 1870 that it would be important to talk between Chicago and New York. Who knows but seeing between New York and Chicago may become as common as telephoning is now? It is a curious fact that we must provide facilities for communication before we can determine how useful they are. Thus it was with the telegraph and the telephone, and thus it will be with television."

Chapter Five

LIFE IS INSTILLED IN THE IMAGES

The dawn of 1927 casting a light on the achievements of the years just passed reveals to the research workers that they are nearer the goal of successful television than ever before. The remarkable advance of radio broadcasting has given them new electric tools to work with in the television laboratories.

And so the race of man to become master over the elusive images that ride through the sky on invisible ribbons of communication becomes more intense and moves at a faster pace.

Photographs Come to Life—April 7, 1927

Like a photograph come to life, Herbert Hoover, Secretary of Commerce, makes a speech in Washington and an audience in New York watches him in action on a screen as they hear him speak. His picture comes to the metropolis by wire at the rate of eighteen images a second so that they appear on the screen as a motion picture. As each syllable is heard the motion of Hoover's lips and the changes in his facial expression flash on the screen.

This is a triumph for television. When the images are about three inches square the likeness is excellent. When the screen is enlarged to two by three feet, the results are not so clear. But, nevertheless, the New Yorkers are thrilled to see the image come to life, as it begins to talk, smile, nod its head and look this way and that. Hoover looks down as he reads his speech, and holds the telephone receiver up so that it covers most of the lower part of his face.

So quick is the transmission that the engineers estimate that the New York hearers and spectators are something like a thousandth part of a second later than the persons at his side in hearing him and in viewing the changes in his countenance. This is all done by wire but the second act in the performance features radio-television between the Whippany, N. J., studio of the American Telephone and Telegraph Company and the New York screen.

The first face to appear on the screen from Whippany is that of E. L. Nelson, an engineer who gives a technical description of what is taking place. He screens well as he talks.

A COMEDIAN APPEARS.—Next is a vaudeville act by television from Whippany. It is an historic performance. A. Dolan first appears. He is a comedian. He does a monologue in brogue. The audience sees him as an Irishman with side whiskers and a broken pipe. Then he disappears. But in a minute he is back on the screen, this time blackfaced with a new line of jokes in negro dialect. It is the first vaudeville act on the air as a talking picture and in its possibilities an observer compares it with the Fred Ott sneeze of more than thirty years ago, the first piece of comedy recorded in the movies.

A short humorous dialect talk by Mrs. H. A. Frederick of Mountain Lakes, N. J., is the next number on the program from the Whippany studio. Before and between the acts an announcer makes a motion picture appearance. He is seen and heard.

Some one recalls that Alexander Graham Bell, the inventor of the telephone, predicted at a meeting in the Times Building, more than twenty years ago, that the day would come when the man at the telephone would be able to see the distant person to whom he was speaking. And now that dream has come true. In the Washington part of this television demonstration a telephone girl is visible. She appears

LIFE IS INSTILLED IN THE IMAGES   67

on the screen and asks to whom the caller wishes to talk. She is a pretty girl with fluffy hair, and it is observed that she is as calm and efficient as if she had been at a television-telephone switchboard all her life.

THE FRUITION OF STUDY.—Walter S. Gifford, president of the American Telephone and Telegraph Company, opens the demonstration with this introduction:

"Today we are to witness another milestone in the conquest of nature by science. We shall see the fruition of years of study on the problem of seeing at a distance as though face to face. The principles underlying television, which are related to the principles involved in electrical transmission of speech, have been known for a long time, but today we shall demonstrate its successful achievement. The elaborateness of the equipment required by the very nature of the undertaking precludes any present possibility of television being available in homes and offices generally. What its practical use may be I shall leave to your imagination. I am confident, however, that in many ways, and in due time, it will be found to add substantially to human comfort and happiness."

The audience realizes that it is to witness an important step in the history of communication. It is recalled to them how on March 10, 1876, Bell stood in a room in a boarding house at 5 Exeter Place, Boston, and spoke into a telephone transmitter, that connected with an adjoining room, to Thomas A. Watson, who had been working with him:

"Mr. Watson, come here. I want you."

Watson came rushing into the room, shouting, "I heard you. I heard what you said."

But even that remarkable invention was neglected until discovered in an inconspicuous corner at the Philadelphia Centennial by Dom Pedro, Emperor of Brazil, and became the sensation of the exhibition. Bell was ridiculed when he predicted that some day it would be possible for men to talk from Boston to New York as easily as from room to room.

And now, in 1927, an audience in New York is seeing Washingtonians by television!

General J. J. Carty steps before the televisor's eyes in Washington and gives the signal for the show to begin. He holds a telephone transmitter in his hand while the light of an arc lamp flickers on his face. Small dots of light are moving across his face, one after another, but at such high speed that they bathe his countenance in uniform illumination that has a bluish tinge. These lights are dissecting his face into small squares. And each tiny part travels over the wire to New York with inconceivable rapidity; in fact, at the rate of 45,000 a second. The receiver reassembles the squares as a mosaic. It takes about 2,500 of the tiny squares—or "units", as they are called—to build up each complete picture. Gifford is at the New York end of the wire to greet Carty.

"How do you do, General? You are looking well," Gifford remarks.

Carty smiles and inquires after the health of the speaker at the New York end.

"We are all ready and waiting here," reports Carty. "Mr. Hoover is here. They are having a little power trouble."

Hoover is called to take a seat so that the light beams can play across his face and send it over the wire to Manhattan Island. In a few seconds the New Yorkers hear his voice and he is seen on the illuminated transparent screen which has a corrugated appearance. This is because the squares which comprise the picture are arranged in fifty rows, one on top of the other.

The room is darkened. At first, in the center of the screen a white glare appears. As the spectators watch the screen they notice that the large luminous patch is forming a forehead—the forehead of Hoover. He is leaning in such a

LIFE IS INSTILLED IN THE IMAGES 69

way that the forehead takes up too much of the picture, while the telephone he is holding blots out the mouth and chin. Then he moves and the picture clears. He is easily recognized.

HOOVER IS TELEVISED.—He looks up from the manuscript, the lips begin to move and this is what Herbert Hoover said in his first television-telephone speech:

"It is a matter of just pride to have a part in this historic occasion. We have long been familiar with the electrical transmission of sound. Today we have, in a sense, the transmission of sight, for the first time in the world's history.

"Human genius has now destroyed the impediment of distance in a new respect, and in a manner hitherto unknown. What its uses may finally be no one can tell, any more than man could foresee in past years the modern developments of the telegraph or telephone. All we can say today is that there has been created a marvelous agency for whatever use the future may find, with full realization that every great and fundamental discovery of the past has been followed by use far beyond the vision of its creator.

"Every school child is aware of the dramatic beginnings of the telegraph and the telephone and the radio, and this evolution in electrical communications has perhaps an importance as vital as any of these.

"This invention again emphasizes a new era in approach to important scientific discovery, of which we have already within the last two months seen another great exhibit—the transatlantic telephone. It is the result of organized, planned and definitely directed scientific research, magnificently coordinated in a cumulative group of highly skilled scientists, loyally supported by a great corporation devoted to the advancement of the art. The intricate processes of this invention could never have been developed under any conditions of isolated individual effort.

"I always find in these occasions a great stimulation to

confidence in the future. If we can be assured a flow of new and revolutionary inventions to maintain thought, stimulate spirit and provide a thousand new opportunities for effort and service, we will have preserved a vital and moving community."

Mrs. Hoover is next invited to sit in front of the televisor and she converses with Mr. Gifford.

"What will you invent next?" she asks. "I hope you won't invent anything that reads our thoughts."

Newspaper reporters then take turns at the televisor. The New Yorkers talk with David Lawrence, a Washington correspondent. In commenting upon the event one reporter said that Lawrence was pictured perfectly on the small screen. He looked like an excellent daguerreotype which had come to life and started to talk. Even the crinkle of his hair registered perfectly. In these small motion pictures, projected by television, the detail of the face appears in clear-cut black lines against a shining gold background, due to the orange light from the neon tube.

DESCRIPTION OF THE PROCESS.—It remains for Dr. Ives to describe the process. Aside from the terrific speed of transmission and the fact that an error of ninety-thousandths of a second in the synchronization between the apparatus in Washington and that in New York would jumble the picture, he assures the audience that the problem is not as complicated as it might seem.

"The performance begins when the person to be televised takes the seat in front of the television eyes," said Ives. "Then the arc light is turned on. The revolving disk shuts most of the light off from the sitter. There is a series of holes along the rim of the disk. As the disk whirls, the light flashes through and strikes the person in front of it, through a hole nearest the rim. That spot of light travels across the top of the head. The second hole is not as close to the rim of the disk. Therefore, the second spot of light travels across the

LIFE IS INSTILLED IN THE IMAGES 71

face, just below the first, and the third just below the second, and so on. There is a total of fifty holes so that fifty spots of light one beneath the other speed across the scene or object to be televised.

"If the process could be slowed down infinitely, it would begin with the action of the visible spot of light. But in actual operation, the spots move so quickly that the subject is flooded by steady illumination. However, there is never more than one spot of light on the face or scene at a time, but the entire fifty spots or daubs of light flash across the face eighteen times in a second. The lines, contours, shadows, highlights and colors of the face naturally cause variations in the brightness of the light spots they reflect. These variations are converted into variations in electric current.

"Three large photoelectric cells face the person being televised. The moving spots of light are reflected from the face into these cells, where they cause an electron shower or flow of electricity. The showers are strong or weak, as the light is strong or weak. These electron showers are nothing but electric current, so that the photoelectric cells cause a current, which constantly varies according to the characteristics of the countenance or scene to be televised. Then the vacuum tube amplifiers are put to work to intensify the current 5,000,000 to 10,000,000 times before it is strong enough to perform the work required of it. Then it is sent by either wire or radio to the receiving set and television screen."

PORTRAITS THAT FLY.—This electricity is literally a flying picture. Every change in volume is the feature of a face or scene. The trick is to make every bit of the flying portrait land in the right place. When it arrives at the receiving station the current is carried to a "brush" or an electrical contact device which is mounted on a wheel. As the "brush" revolves on the wheel it makes and breaks the electrical contact approximately 2,500 times. Each contact is made with one of 2,500 wires or "nerves" mounted on a circle in which the wheel whirls. Each "nerve" snatches a bit of the electric current or flying picture. This wheel must spin so accurately, in synchronization with the revolving disk at the transmitter, that each one of the "nerves" will have delivered to it eighteen times in a second exactly the bit of picture intended for it. The slightest error would scramble the portrait.

Each wire-nerve carries its bit of current to a square of tinfoil behind the television screen. These patches of tinfoil are arranged fifty in a row. And there are fifty rows. When the bit of current carrying a tiny fragment of the picture reaches one square of tinfoil, it leaps to a wire. It makes the jump through a bulb in which there is neon gas. The glow of this lamp is instantaneously effected by the passage of the electric current through it. Eighteen times every second there is a flash of light in front of each of the 2,500 patches of tinfoil. The flash is strong or feeble, according to the light or shadow on a particular part of the face or scene. These rapid flashes build up the picture of the screen and they do it at the rate of about 45,000 flashes a second.

The transmitting and receiving of the picture, that is, taking it to pieces at one place and reassembling at another, is synchronized by a special method which causes every one of the 2,500 squares or picture units to fall in the proper place eighteen times a second. This control calls for the use of two wires. And in the case of radio television, one wave length is employed for sending the picture and two others for the synchronization process. This necessity for at least three wave lengths is an obstacle in the path of sending television to the home, because the radio lanes are already badly congested, except in the ultra-short wave spectrum.

Ives emphasizes that it would require several hundred times as many dots of light, under equally perfect control, to make television practical on a large screen such as utilized in motion picture theaters. Furthermore, television cannot be thrown on a larger screen without the use of a more powerful flood of illumination on the person or scene televised. The light used in this experiment is strong enough to be uncomfortable to a person sitting within its glare for any length of time. However, this factor is not likely to remain as a heavy obstacle. It is expected that a more sensitive photoelectric cell will be developed which will enable the television camera to function in less intense illumination.

FACES KNOWN BY THEIR SOUNDS—APRIL 8, 1927

Strange as it may seem, television signals can be heard as well as seen. Any radio receiving set if in tune with the proper wave can eavesdrop on the show. But if only earphones or a loudspeaker is utilized, television is a mere squeal of varying intensity. One can chalk $1,000,000 on a blackboard and when held in front of a televisor it will sound far different than a dollar bill within range of the electric eyes. Every scene and every object has a characteristic tone or squeal.

John Baird makes a trip to Glasgow to show the homefolks what he is doing with television. He shows them that every face has a characteristic sound. A blonde sounds different from a brunette. Even a derby hat, whether it be brown or black, has a different refrain than a cap or gray felt hat. A Scotch plaid sounds as distinctive as it looks. In fact, every substance emits a distinctive refrain when televised and picked up by a loudspeaker instead of on the screen.

THE BAIRD DEFINITION.—"Television may be defined as the transmission by telegraphy of images of actual scenes with such rapidity that they appear instantaneously to the eye," said the Scotsman at the Glasgow meeting. "The eye, unfortunately for the success of television, has a time lag, and images therefore need not be transmitted instantaneously. The problem of television has been approached by two different methods. The first and most obvious was to build apparatus in imitation of the human optical system. The human eye consists essentially of a lens which casts an image of the object viewed upon the retina. The surface of the retina consists of several million hexagonal cells into which lead the ends of the optic nerve. These nerve terminals are immersed in a light-sensitive substance, the visual purple, which, when ionized by light, changes its color from purple to a grayish yellow.

"This ionization of the visual purple sends impulses along the nerve fibers to the brain. The visual purple in life is continually renewed, so that, in effect, it might be compared to a motion picture camera, with this difference, that in place of using a moving film coated with a light-sensitive emulsion, the light-sensitive visual purple is continually changed."

Baird, like other inventors in describing their work, asks the audience to keep in mind that the eyes are a human television system. The scenes they view are transmitted to the brain as mosaics comprising an enormous number of tiny areas, each of which is flashed simultaneously to the receiving centers in the brain. There the impulses of the optic nerve produce mosaics corresponding to the images on the retina. He says that artificial television models on these lines were actually suggested by several early experimenters, but they soon discovered that the stupendous number of cells, wires and shutters required made the development of such a scheme out of the question.

THE SECOND APPROACH.—Baird then describes the second method of approach to television. It uses one photoelectric cell and causes each of the elemental areas to fall in quick succession upon that artificial eye.

"About four years ago I decided to devote my entire time to achieving television," he continued. "The problem seemed comparatively simple. Two optical exploring devices rotating in synchronism, a light-sensitive cell and a controlled, varying light source capable of rapid variation were all that was required. They appeared to be already known. The problem of synchronism had apparently been solved in a practical way in multiplex telegraphy. Quite a number of optical exploring devices were available. The photoelectric cell, in conjunction with the thermionic valve (vacuum tube), appeared to offer a ready-made light-sensitive device, and the glow of a discharge lamp an ideal light source. I wondered why in spite of the apparent simplicity of the task none had produced television. I found that the stumbling block was in the cell. After six months' work, however, I managed to get shadows through. Then the step from mere shadows to images by reflected light proved extremely difficult, but in April, 1925, I had the satisfaction of transmitting simple outlines."

The inventor pauses here in his lecture to demonstrate by means of a loudspeaker diaphragm how each object that is televised sings its own refrain.

"If television transmissions are received on a telephone," he explains, "they are audible, because every object or scene has a characteristic sound. I have made a few phonograph records of the sounds created by different persons' faces. By noticing carefully it is possible to distinguish one face from another. A further interesting point is that these records can be turned back into images, so that a living scene can be stored in the form of a phonograph disk."

In Baird's first demonstrations of television he had to utilize an intensely brilliant illumination, which caused considerable discomfort to the person being televised. By using infra-red rays, however, he ultimately dispensed altogether with light, with the somewhat remarkable result in that the television eye could see in total darkness.

Part III

A NEW DECADE IN RADIO VISION

OPPORTUNITIES

. . . any one who has witnessed the new inventions, the birth of new industries, the acceleration of production and consumption, and the structural changes which have so vastly increased the wealth of the world and altered our entire mode of living within the memory of those present, cannot be discouraged about either the immediate or the distant future.

The opportunities which have so multiplied in the last generation are only the forerunners of others, and perhaps greater ones, which will come as the result of forces now at work and constantly being discovered, so that it is impossible to predict what may be the opportunities that lie immediately ahead. . . .

ANDREW W. MELLON,
Secretary of the Treasury, at American Bankers' Association Convention, 1931.

Chapter Six

SEEING ACROSS THE ATLANTIC

Boyhood cultivates the restless ambition for youth to go to sea. So it is with science, ever youthful, ever adventurous, ever seeking new realms in which to conquer. So it is with wireless.

After Marconi had nurtured his invention past the baby age it suddenly left the land and darted out into the emptiness of space over the ocean to seek a slender target on the distant shore. And television, after it peered through the London fog and across the foothills of the Adirondacks, wanted to glimpse farther. It wanted to see across the Atlantic!

That great expanse of water has beckoned many an adventurer to fame and many others to disaster since Columbus first accepted its challenge. The Atlantic brought laurels to Marconi. It brought glory to Lindbergh. It gave prestige to the Zeppelin. It gave the submarine a mystic power. It is good to all who triumph on its waves or in the sky above them. It is cruel to those brave souls struck down by Fate in man's battle to annihilate time and space.

To look beyond the sea was a natural ambition for television images, anxious for a longer flight that would prove beyond all doubt the power of electric eyes to see far across the horizon of the earth.

It is 1928!

An Image Crosses the Sea—February 8, 1928

It is a cold night. The air is crisp and the stars are twinkling in a clear winter sky. Twenty-seven years have gone since the letter "S" made history in its flight across the Atlantic as the first wireless signal to leap that distance. The ever restless ocean has been further conquered since then. Millions of dots and dashes, thousands and thousands of spoken words have sped across the three thousand miles of water. Airplanes and dirigibles have soared high out of range of King Neptune's pronged fork and have landed safely on the other side of the sea.

Tonight science is engaged in another contest with the Atlantic. Again the great expanse is to be defied. Over in England, Big Ben struck midnight several hours ago. A number of radio experimenters are neglecting their slumber to tune up electrical apparatus and to adjust a new kind of man-made eye. Far over the horizon, not far from New York, another group hovers around a receiving set, the currents of which flow into a unique contraption that resembles a lens. On the roof a slender wire dangles between two masts always ready to pluck energy from passing radio waves no matter from where they may come.

Mrs. Mia Howe sits under the glare and heat of the powerful electric lamps in the laboratory of John Baird in London. In front of her is a black wall in which there is an opening about a foot square. It is the gateway to a scanning disk. The inventor calls through a speaking tube as he focuses the apparatus, "Face up a little closer. Chin up, please."

Through the hole in the wall Mrs. Howe sees a great wheel going round and round. A slotted disk whirling 2,000 revolutions in a minute interrupts the light and reflects the image, causing it to reach the light-sensitive cell in a series of flashes. She describes it, saying that it looks like a sawmill, but to Baird it is part of a machine that is sending the sound of a face across the ocean.

Over in the United States in a darkened cellar in the village of Hartsdale, N. Y., a group of persons watch Mrs.

SEEING ACROSS THE ATLANTIC

Howe turn her head and move from side to side. The images are imperfect, but they are images nevertheless. Transatlantic television is a reality! Another dream of science is on the way to realization.

The face crosses the sea as a rhythmic rumble, at least that is the way it sounds as it hums like a bumblebee while being transformed into a picture by a little black box. The musical buzz and its choppy cadence represent the lights and shadows of the face. That is the way a television signal sounds when tuned in by an operator wearing earphones. And that is the way it sounds if by chance it goes through a loudspeaker.

The black box is the televisor. It has a gaping eye in which tiny oblongs of light suspended in a whirling rectangle of brilliance swirl and shift to form the face of the woman far across the swells of the Atlantic. The elements that lurk in the air over the sea this night try to "break up the show." The face of Mrs. Howe appears broken and scattered, but those who see it have no doubt that it is a woman, as she first shows full face and then her profile.

Then some one pushes a Jack-o'-lantern in front of the televisor in London. The Americans see it turn its head from side to side and open its mouth. It is the first Jack-o'-lantern to pose for a transoceanic broadcast.

The receiving is done by R. M. Hart, owner of a short wave radio station using the call 2CVJ. Two kilowatts of power are employed to lift the face across the ocean waves.

Trying to Rival Nature—February 10, 1928

A London surgeon has been persuaded to give John Baird the eye just removed from a boy, in order that he might try it in his television machine in an effort to rival nature.

"As soon as I was given the eye," said Baird, "I hurried in a taxicab to the laboratory. Within a few minutes I had the eye in the machine. Then I turned on the current and the waves carrying television were broadcast from the aerial. The essential image for television passed through the eye within half an hour after the operation. On the following day the sensitiveness of the eye's visual nerve was gone. The optic was dead. Nothing was gained from the experiment. It was gruesome and a waste of time.

"I had been dissatisfied with the old-fashioned selenium cell and lens. I felt that television demanded something more refined. The most sensitive optical substance known is the nerve of the human eye. It was essential to get some visual purple in the natural setting of the human eyeball in order to use it as a standard of perfection in completing the visual parts of my apparatus. I had to wait a long time to get the eye because unimpaired ones are not often removed by surgeons."

BAIRD'S ROAD TO TELEVISION—FEBRUARY 19, 1928

Captain O. G. Hutchinson, an aide of John Baird, has arrived in New York to supervise tests which he hopes will lead to transoceanic broadcasts that will enable New Yorkers and Londoners to see each other no matter how thick the fog that might be hovering over the Thames or Sandy Hook.

He is in a reminiscent mood. He refers to Baird as "the Galileo of radio vision."

"When but twelve years of age Baird began making selenium light-sensitive cells to transmit shadow pictures, the forerunner of his later televisor," said Hutchinson. "He often burned his fingers so badly in his early work with chemicals that his father, a minister of West Parish church, Helensburgh, Scotland, frequently felt the necessity of apologizing to his friends for the condition of his son.

"Undismayed by the predictions of older scientific minds in England, made less than three years ago, that twenty-five to fifty years would have to elapse before Baird's 'child' would crawl from the laboratory, he worked on in the direst poverty and under the most adverse and squalid conditions. Late in 1925, I happened to meet him on the Strand in London with patches on his clothes and about ten dollars in his possession. That was all he had left of a half interest in his invention which he had sold to a friend for money to keep him alive and to carry on his experiments.

"In December, 1925, we undertook to interest some friends in the possibilities of the invention and succeeded in raising a few thousand pounds when Baird said he could produce television apparatus for a demonstration in six months. That marked the beginning of the upward trend and recognition among those who had predicted failure.

"Baird had his first position with the Argyle Motor Works in Alexandria, Scotland, where he worked in the drawing office," continued Hutchinson. "From there he went to the Clyde Valley Power Company, near Glasgow. During the World War he invented what is known in England as the 'Baird Under-sock,' which was worn by soldiers in the trenches to prevent or allay the malady called 'trench-feet.' He has most eccentric habits. Upon occasions when he wants to think intently he goes to bed for a week at a time. He said that upon one of these prolonged rests he conceived the working part of his apparatus."

Hutchinson remembers one evening while sitting on the roof of the laboratory in London, Baird commented on the blood-red sunset, saying that the deep color was caused by the red rays, which, being of longer wave length, were able to penetrate the London mist with greater facility than the other colors. He wondered why the invisible rays just beyond the red, known as infra-red rays, could not be used to bring about sight in darkness. They might be able to penetrate mists and interfering media better than the red visible rays.

He went into the laboratory for a week. At the end of that

time he told friends he had an interesting experiment. He invited them to enter the darkened laboratory. The apparatus was switched on and the guests beheld an image on the screen, the likeness of a manikin at the other end of the room.

As a further test an experimental image was placed inside a glass case in which a chemical fog was so dense that the image could not be seen. The seeing-in-darkness apparatus, or the "noctovisor," as it is called, penetrated the fog by means of the infra-red rays, which are just beyond the range of human vision, and the image appeared.

"As the next step," said Hutchinson, "talking films were made by means of equipment that evolved from the 'noctovisor' experiments. Baird was discouraged in this venture when existing patents along this line were discovered to be the property of Ernest Ruhmer, a German inventor. This experience caused Baird to turn attention more completely to television, and he succeeded in making his first workable apparatus and demonstrating it in the form of shadowgraphs at the Selfridge store in London in April, 1925. These graphs, however, were a thousand miles from television as we know it today. His main problem was to find a cell that would be sensitive to weak reflected light."

A Face Is Picked Up at Sea—March 7, 1928

Chief Radio Operator Stanley Brown is on board the S.S. *Berengaria*. Miss Dora Selvy is in London, a thousand miles away. For twenty minutes she sits in front of the big eyes of Baird's television station in London, while Brown in his wireless cabin on shipboard watches her smile as her brown eyes look straight at him from the television receiver of the *Berengaria*.

Brown recognizes her quickly because of a characteristic little habit of arranging her dark bobbed hair at the back of her head. And when Miss Selvy turns and appears in profile he is convinced of her identity. She talks, smiles and turns around just to prove to him that the image is that of a living being and not a photograph. She is excited, especially, when a wireless to the ship reports success of the experiment. And then she asks, "I wonder how I looked so far away?"

STEPPING OUT OF THE LABORATORY—JULY 12, 1928

A radio camera is wheeled out on the roof of the Bell Telephone Laboratories. A man dressed in white flannels and sport shirt stands about twenty feet away. He whirls a tennis racket through all the strokes in a tennis player's repertory while the electric eye observes him executing lofts and lobs. Three floors below several persons watch every handstroke he makes. Television is no longer limited to catching the mere head and shoulders of a man sitting in a radio studio. It is on its way to carrying panoramas, spectacles and even mob scenes across miles of atmosphere.

The engineers explain that the trick of recording the action of the tennis player is accomplished by admitting sunlight into partnership with radio. The glaring lamps hitherto necessary in photographing an image have given way to the rays of the sun. A more sensitive photoelectric cell makes this development possible. It will work even on a cloudy day.

"We can take this camera machine to Niagara, to the Polo Grounds or to the Yale Bowl," said Paul B. Findley, of the Bell Laboratories, "and it will pick up the scene for broadcasting. The important step we have taken in this new development is that sunlight is used instead of a powerful artificial light. It will 'photograph' the cataract of Niagara. We could mount the televisor camera on a platform and revolving tripod at a prize ring and broadcast the fight scene. Television has stepped out of the laboratory as far as transmission is concerned. We are no longer limited to studio work."

The television camera utilizes a cloth hood in much the same fashion as did the old style of camera. A lens five inches in diameter projects from an opening in the black cloth. Behind the lens and hidden by the cloth is Nipkow's disk, measuring three feet in diameter. It has fifty tiny holes along its outer rim. They measure one-sixteenth of an inch in diameter and are so arranged that no line effect is noticed on the picture at the receiving end. The disk is of aluminum, painted black, and when in operation it revolves so that each of the little light openings passes a given point eighteen times in a second. This creates 900 lines to "paint" the image. The impulses are so rapid that the lines are blended and the picture looks like a half-tone.

The great sensitivity of the new photoelectric cell is the reason why less light is required on the image. The camera will even operate on a hazy day, but clouds that shut off the sun, naturally, lower the efficiency of the machine.

CAMERA SUPPLEMENTS MICROPHONE.—The image is focused by moving the lens back and forth. In fact, the principle is the same as that of a kodak. In the first form of equipment demonstrated in April, 1927, the scene was illuminated by a rapidly oscillating beam from a powerful arc light. The scene to be broadcast was thus limited to a small area. The new machine frees television from this serious limitation. This experiment shows that persons in motion and objects a considerable distance away from the camera can be scanned successfully.

Dr. Frank Gray is in charge of the camera during the demonstration. He explains that the radio camera can be employed several miles from the broadcasting station and be connected to it by wire line, as are microphones that pick up music and voices at points outside the studio.

The receiver is shrouded in darkness. One merely sees a picture about two and a half inches square of the tennis player on the roof as he jumps about and swings his racquet. The engineers, however, assure the audience that the receiver represents no new development.

"This demonstration is merely to illustrate another advance," said one of the engineers. "It does not mean that television will be ready for use in every home equipped with radio tomorrow. The equipment is too elaborate for home use. It requires experts to operate the instruments, although part of the receiving station is an ordinary broadcast receiver. We hope to make the pictures larger. We hope to have television so that it can be used in the living room without having the room darkened. Perhaps some day we will flash the images on a screen like the movies, but when that will be we cannot say."

CURTAINS ARE DRAWN BACK—SEPTEMBER 11, 1928

Science is pushing asunder still further the curtains behind which man catches a glimpse of nature's secrets on the stage of Time. The genial Dr. Alexanderson in the rôle of a scientific showman gives an audience at Schenectady a glimpse of what may be expected on a more spectacular scale in the future.

For the first time in history, a dramatic performance is broadcast simultaneously by radio and television. Voice and action travel together through space in perfect synchronization, in a forty-minute broadcast of J. Hartley Manners' one-act play, *The Queen's Messenger*. It is an old spy melodrama, for years a favorite with amateur Thespians. It is chosen for this experiment because the cast contains only two actors, and their parts are such that they can alternate in front of the television camera.

While the actors play their rôles in a locked studio, the audience sees and hears them through a television receiving set in another room in the same building. Their appearance and voices, translated into electrical impulses, are carried by land wire to the broadcast transmitter of WGY four miles away, where they are given wings for a flight through space. They are picked up again at the place of their origin. The effect is the same as listening to a radio drama, except that in addition to hearing the sounds the audience sees moving pictures of the actors as they speak their lines and do their stage "business" with cigarettes, cigars, knives, pistols and other "props."

The pictures are about the size of a postal card and are sometimes blurred and confused. They are not always in the center of the receiving screen. Sometimes they are hard on the eyes because they flicker. It is recalled, however, that ordinary moving pictures suffered from worse mechanical defects in their pioneer days, and that these shortcomings of the television pictures signify that they are still in the experimental stage.

THE INVENTOR'S PREDICTION.—Alexanderson makes it clear that it will be some time before radio vision is perfected to such a degree as to make it practical for home entertainment. He predicts that some day there will be special television theaters all over the world, without actors, musicians, scene shifters or stage hands, receiving simultaneously identical theatrical broadcasts and musical performances from a central broadcasting station. And some day, he believes, television will be seen in natural colors. He is already at work on the next step, which will give a performance with pictures measuring a foot square and later full-size motion pictures of the head and face.

He points out that as far as the main demonstration featuring the one-act play is concerned, the most significant factor is the synchronization of word and gesture. That is a step forward. And he declares that great as has been the triumph of the talking movies, they may easily be outdone

SEEING ACROSS THE ATLANTIC

by television if the technical difficulties are overcome, because then radio will carry both words and scenes of dramatic and musical performances, besides public events and athletic games, into the homes as well as into the theaters.

Like many of the early sound broadcasts, *The Queen's Messenger* is released into the air by radio, but how far it travels no one knows. There are no telegrams, telephone calls or letters to tell of success. It is broadcast at 1:30 o'clock in the afternoon and repeated at 11:30 P.M., in hopes that under the cover of darkness the waves might go across a greater mileage to be intercepted by amateur television experimenters. Several amateurs along the Pacific coast have picked up images broadcast from the Mohawk Valley on previous occasions, but they regard themselves as lucky when able to hold one of the elusive faces for thirty seconds on the screen.

The two characters in the cast of *The Queen's Messenger* are played by Izetta Jewell, a former star of the legitimate stage, the wife of Professor Hugh Miller of Union College, and Maurice Randall of the WGY Studio Players. They will be remembered as pioneer television actors.

The technique of handling this television drama makes it necessary for the actors to have two assistants, Joyce Evans Rector and William J. Toniski, whose hands "double" for the hands of Miss Jewell and Mr. Randall in certain scenes. They are needed to manipulate the "props" before the television camera. This is necessary because, at the present stage of progress, only the heads of the two actors can be televised and broadcast.

There are three cameras on duty in the studio. One takes only the scenes in which Miss Jewell appears, another only the scenes which feature Randall, and the third only the scenes in which the hands of one or the other or the various "props" are displayed. In addition to the cameras a microphone functions in front of each player. The director of the

90 THE OUTLOOK FOR TELEVISION

production operates a control box in an effort to bring each actor on the screen at the proper time and to "fade" the actors in and out of each scene as is done in the movies. In front of the director is a special receiving set that enables him to see the faces as they appear on the television screen and check them with the voices.

Three different wave lengths carry the performance; the pictures travel on 379.5- and 21.4-meter waves while the words use the 31.96-meter channel.

Chapter Seven

TELEVISION IN NATURAL HUES

The divine power of the human eye enables man to distinguish colors.

Colors are delicate to reproduce. So man in his work to emulate nature or to photograph its beauty has in many instances found it necessary to be satisfied with black, white and shadows. He could photograph in black and white long before the camera achieved ability to register color on a plate or film. Even today colored motion pictures are special screen productions.

But some day cameras will photograph the colors of the sunset as easily as they do black and white. The "talkies" will be in natural hues. Ultimately television in the home will be in color. When the cataract of Niagara flashes on the screens in millions of homes scattered throughout the world that beautiful deep green of the tumbling tons of water will be seen in vivid contrast with the white, madly tossing water of the rapids, just before the dash over the precipice.

Such will be the wonder of seeing by radio, but first man must be satisfied with the scenes in black and white, because science finds it a Herculean task to match the power of the human eye.

It is 1929!

Television Blossoms in Color—June 27, 1929

The Stars and Stripes fluttering in front of watchful electric "eyes" in the Bell Telephone Laboratories is reproduced in red, white and blue on a screen about one hundred feet from the transmitter to show the possibilities of television in color. In this experiment a radio channel is not used. Wires link the transmitter and receiver. The principle is the same in either case.

The observer walks into a darkened booth and through a peek hole sees the American flag in color. It is about the size of a postage stamp. The colors reproduce perfectly. Then the Union Jack waves on the screen and is easily recognized by its colored bars.

The man at the transmitter in this television theater, the technical impresario of which is Dr. Ives, picks up a piece of water melon, and there can be no mistake in identifying what he is eating. The red of the melon, the black seeds and the green rind look true to nature, and so does the red of his lips, the natural color of his skin and the black hair.

Then a pot of geraniums appears as proof that television can reproduce flowers—the red blossoms and the green leaves. Next a large ball with colored stripes; a pineapple, a bouquet of varicolored roses and the image of a young woman in a plaid dress flash on the screen to give evidence that the latest radio "eye" is sensitive to any color.

Ives calls attention to the fact that the engineers have succeeded in adding color to television motion pictures without sacrifice of detail. The picture is restricted to the size of a postage stamp so that not even the finest detail is lost. Once this is achieved, the research experts say it will not be difficult to flash radio movies in color on a much larger screen, although it will be more expensive.

The person or object televised is rapidly scanned by a beam of flickering bright light, while three sets of electric eyes (photoelectric cells) are arranged to transmit current corresponding to the amount of a primary color, red, blue or green. Then at the receiver three tubes form images corresponding in brightness and color to what the electric eyes at the sending machine see. A system of mirrors combines the three images to form the one image in color on the screen.

SIGNIFICANT FEATURES.—One of the most significant features of this color-television achievement is that it does not require complete new apparatus. Much of it is the same as employed in handling television pictures in black and white. The same light sources, driving motors, scanning disks and synchronizing systems and the same type of amplification are used. The only new features are the type and arrangement of the photoelectric cells at the sending end and the neon and argon lamps at the receiver.

The outstanding contributions that have made colored television possible are a new photoelectric cell, new gas cells for reproducing the image and the instruments associated directly with them. To render the correct tone of colored objects it was necessary to obtain photoelectric cells, which like the modern orthochromatic or panchromatic plate, would be sensitive throughout the visible spectrum. This requirement has been satisfactorily met. Through the work of A. R. Olpin and G. R. Stilwell a new kind of photoelectric cell has been developed which uses sodium in place of potassium. Its active surface is sensitized by a complicated process using sulphur vapor and oxygen instead of a glow discharge of hydrogen, as with the former type of cell. And the response of the new cell to color, instead of stopping in the blue-green region, continues all the way to the deep red.

Each of the three groups of photoelectric cells is provided with color filters or sheets of colored gelatine. One has filters of an orange-red color, which makes the electric eyes "see" things as the sensitive nerves of the retina sees them. Another has yellow-green filters to give the green effect and the third is a greenish-blue filter which performs a corresponding duty for the blue constituent of vision.

BLENDING THE COLORS.—The former potassium cells were responsive only to the blue end of the spectrum; therefore,

94 THE OUTLOOK FOR TELEVISION

objects of a yellowish color appeared darker than they should have and the tone of the reproduced scene was not quite correct. This disadvantage applied particularly to persons of dark or tanned complexion. When the new cells are used in the original television apparatus and with yellow filters—similar to those used in photographing landscapes in order to make the blue sky appear properly dark—this defect is corrected and the images assume their correct values of light and shade no matter what the color of the object or the complexion of the person. The new photoelectric cells make color television possible.

The development of color television has been greatly simplified by the fact that as far as the eye is concerned any color may be represented by the proper mixture of just three fundamental colors—red, green, blue. This fact is utilized in the development of color photography, all the research of which serves as the background for color television. Several methods of combining the three basic colors to form the reproduced image are available, but, in so far as the sending or scanning end is concerned, the method developed has no counterpart in color photography.

MORE "EYES" ARE UTILIZED.—The scanning disk and the light source are the same as with the beam scanning arrangement used in monochromatic television. The only difference is in the photoelectric cells, according to Ives. And thanks to the trichromatic nature of color vision, it is only necessary to have three times the number of cells used previously, to reproduce all colors. Three series of television signals, one for each set of cells, are generated instead of one and three channels are used for the transmission of the television signal.

The photoelectric cell container is called a cage. Twenty-four cells are located in it. Two have blue filters, eight have green filters, and fourteen are equipped with red ones. These numbers are so chosen with respect to the relative sensitiveness of the cells to different colors. The cells are placed in three banks, one bank in front of and above the position of the scanned object, one bank diagonally to the right, so that the cells receive light from both sides of the object and above. In placing the cells they are so distributed by color as to give no predominance in any direction to any color. In addition, large sheets of rough, pressed glass are set up some distance in front of the cell cage so that the light reflected from the object to the cells is well diffused.

"The receiving apparatus consists of one of the sixteen-inch television disks used in our earlier experimental work," said Ives. "Behind it are the three special lamps and a lens system which focuses the light into a small aperture in front of the disk. The observer, looking into the aperture, receives, through each hole of the disk as it passes by, light from the three lamps—each controlled by its appropriate signal from the sending end. When the intensities of the three images are properly adjusted he therefore sees an image in its true colors, and with the general appearance of a small colored motion picture."

Color television constitutes a definite further step in the solution of the many problems presented in the electrical transmission of images. It is, however, obviously more expensive as well as more difficult than the earlier monochromatic form, involving extra communication channels and additional apparatus. The great obstacle in the way of applying colored television to radio is that it requires so much space in the ether—three invisible channels 20,000 cycles wide. Some method must be found to whittle this to a narrower pathway through the sky.

TRIO OF IMAGES MUST APPEAR.—"For color television the three images must be received in their appropriate colors and viewed simultaneously and in superposition," said Ives. "The first problem was to find light sources which, like the neon lamp previously used, would respond with the requisite fidelity to the short-wave signals of television, and at the same time give red, green and blue light. And when such lamps were available, a decision had to be made as to how the three colors could best be combined to form a single image.

"Thus far the images have been received in a manner similar essentially to the method of monochromatic television. The surface of the disk similar to that used at the sending end is viewed and light from the receiving lamp is focused on the pupil of the observer's eye by suitable lenses. To combine the light of the three lamps, they are placed at some distance behind the scanning disk and two semi-transparent mirrors are set up at right angles to each other but each at 45 degrees to the line of sight. One lamp is then viewed directly through both mirrors and one lamp is seen by reflection from each.

ARGON TUBES ARE USED.—"The matter of suitable lamps to provide the red, green and blue light has required a great deal of study. There is no difficulty about the red light because the neon glow lamp which has been used previously in television can be transformed into a suitable red light by interposing a red filter. For the source of green and blue light nothing nearly so efficient as the neon lamp was available. The decision finally made was to use another one of the noble gases—argon—which has a considerable number of emission lines in the blue and green regions of the spectrum. Two argon lamps are employed, one with a blue filter to transmit the blue lines and one with a green filter transparent to the green lines of its spectrum.

"These argon lamps unfortunately are not as bright as neon lamps; therefore, it was necessary to use various expedients to increase their effective brilliancy. Special lamps to work at high-current densities were constructed with long narrow and hollow cathodes so that streams of water could cool them. The cathode is viewed end-on. This greatly foreshortens the thin glowing layer of gas and thus increases its apparent brightness. Even so it is necessary to operate these lamps from a special tube amplifier to obtain currents as high as 200 milliamperes."

It is easily understood that television in colors is a far more difficult task than is monochromatic television. Errors of quality which would pass unnoticed in an image of only one color may be fatal to true color reproduction where three such images are superimposed and viewed simultaneously. In three-color television any deviations from correct tone throw out the balance of the colors so that while the three images might be adjusted to give certain colors properly, others would suffer from excess or deficiency of certain of the constituents. A further source of erroneous color exists at the scanning end. If the light from the object being televised were not distributed equally to all the cells, the object would appear as if illuminated by lights of different colors shining on it from different directions.

The Kinescope Is Introduced—November 18, 1929

There are no moving or mechanical parts in a new television receiver that Vladimir Zworykin displayed at the district convention of the Institute of Radio Engineers at Rochester, N. Y. The image appears on the flat end of a cone-shaped cathode ray tube. It takes the place of the neon or glow tube, the scanning disk and the motor of previous television receivers. It is noiseless in operation and there is practically no difficulty in synchronizing the transmitter and receiver. Zworykin, who has apparently discovered several missing links in television, calls the new tube a "kinescope." He thinks that eventually this improvement will mean practical simplified television for the home.

This cathode ray tube produces a picture with less flicker than does the scanning disk. The image is four by five inches, but the inventor is confident that additional experiments will teach him how to build tubes which can produce larger pictures. To make it possible for a number of people to watch the images the active surface of the cathode tube is located below a tilted mirror which reflects the action and permits several to see it at the same time.

The big feature of this system is that a receiver has been developed without complexities such as the whirling disk which must be always in exact step with the scanner at the transmitting end.

Speaking of television or the projection of motion pictures by radio, Zworykin says: "All the processes needed for projecting motion pictures are in existence. The theory is all right but at present the apparatus would have to be endless, cumbersome and uncertain. But it will be simplified. It will take some years, but we will have eventually the instantaneous or near-instantaneous transmission of sound motion pictures by radio. I am ready to discuss the practical possibility of flashing radio images on motion picture screens so that large audiences can view the television broadcasts of important events as sent out from a central station. Visual broadcasts in the future will be synchronized with sound."

The Zworykin machine is based on the principle that a pencil of electrons from the hot cathode bombards a screen of fluorescent material which glows where the electrons strike it. The electronic pencil follows the movement of the scanning light beam so rapidly that the eye beholds a perfect impression of a miniature motion picture.

The transmitter comprises a motion-picture projector rebuilt so that the film passes the film-gate downward at a constant speed. This movement is the vertical component of scanning. The horizontal scanning is accomplished by sweeping the film crosswise with a point of light traveling more rapidly than the downward movement. As a result the picture passing through the gate is scanned crosswise and from top to bottom by a series of horizontal lines of light.

The illumination is supplied by an ordinary automobile

bulb focused on a diaphragm on the projector. This in turn is focused on the film and the light which passes through it is again refocused in the form of a stationary spot that strikes the photoelectric cell.

DUTY OF THE TUBE.—The function of the cathode ray tube in the television receiver is twofold: first, it converts the electrical impulses received from the transmitting station into light impulses. Since the electrical impulses from the transmitter represent the variation of light intensity of the transmitted image, the light variation obtained on the screen of the cathode ray tube reproduces the image.

The second function of the cathode ray tube is to reproduce the scanning of the image without the use of moving mechanical parts. When the transmission is accomplished by means of a Nipkow disk, the image is scanned by a series of parallel lines which cover the whole area of the image. Exactly the same set of parallel lines is reproduced on the screen by the cathode ray beam by deflecting it with magnetic or electrostatic fields. These fields, of course, are so adjusted that the movement of the spot across the screen follows exactly the movement of the scanning spot of the transmitter.

The duty of the cathode ray tube, when used for transmitting purposes, is entirely different from that in the receiver and, therefore, the design of the transmitting cathode ray tube is entirely different from the receiving tube.

ADVANTAGES OF CATHODE RAY TUBE.—Zworykin sums up the advantages of the cathode tube in the television receiver as follows: No mechanical moving parts are used. Therefore, the set is more easily operated by the rank and file of the radio audience. It is quiet in operation. Synchronization of the transmitter and receiver is easily accomplished, even when a single carrier wave is used. There is ample amount of light. The persistence of fluorescence of the screen aids the persistence of the eye's vision. Therefore, it is possible to reduce the number of picture units per second without any flickering effect. This in turn allows a greater number of scanning lines and consequently a picture of greater detail without increasing the width of the radio channel. The light and electron beams having no physical weight compared with moving mechanical parts offer no resistance to the device utilized to gain accurate synchronous operation of the transmitter and receiver.

CROOKES DISCOVERS THE RAYS.—Cathode rays, the luminous streaks of which paint the television picture, were first discovered by Sir William Crookes in the 'eighties. The tube in which the rays perform is funnel-shaped. The wide end is sealed and the slightly convex cover is coated with a fluorescent material (Willemite or a similar acting substance), behind which is hidden the so-called electron gun that shoots the pencil-like stream of electrons against the fluorescent screen on which the image appears. Constant tests are being made to find fluorescent coatings that will glow with greater brilliancy. The research experts are secretive about this feature of their cathode ray bulbs.

The electrons traveling at high speed make a rapid trip through the tube, excite the gas molecules in their path and the fluorescent screen glows when the electron streams strike it. The electrons are endowed with kinetic energy and momentum because of their great velocity. This cathode ray tube is sometimes called an oscilloscope, in fact, these tubes designed for television are closely related to that electrical instrument known as an oscillograph.

The tube as employed in television has two parallel metallic plates upon which an electrostatic charge can be placed. And there are two coils which produce a magnetic field when an electric current is sent into them. The purpose of these intermediate devices located close to the source of the electron flow is to deflect the beam of electrons either in a vertical or in a horizontal plane. The cathode beam, because it consists of electrons, is sensitive to both magnetic and electrical fields of force. Therefore, when the intensity of either of these fields is altered a spot of light at the end of the bulb is caused to move. It draws a bright fluorescent line as it passes over the "screen" end of the tube. This line can be made to move with such rapidity up and down in lines so close together that the human eye views the end of the tube completely aglow. Now the trick is to obtain a picture from this phenomenon. That is where the magic of Zworykin, Manfred von Ardenne of Germany, Farnsworth and others enters.

First of all, they know that to produce a picture it is necessary to have various intensities of light on the screen. They accomplish this by varying the intensity of the electron beam when it sweeps across the end of the tube. A high intensity electron beam creates a bright area while low intensity gives weak illumination.

To understand this phenomenon it is helpful to recall the analogy of the image of the Indian's head traced on a piece of paper under which is a coin. As the pencil moves across the paper the raised part of the coin stands out in relief while the background is lighter. Difference in pressure gives the result in this case. High speed of the electron pencil in the cathode ray bulb produces a similar effect on the fluorescent screen.

Two Types of Cathode Ray Tubes.—There are two types of cathode ray tubes, "cold" and "hot." The difference between the "cold" and "hot" cathode ray tube is in the method by which the electron stream is produced. In the "cold" tube, it is produced by the discharge through the residual gases, and in the "hot" tube it is emitted from the electrically heated filament. Only the latter type tube is used for television purposes.

The kinescope is of the "hot" cathode variety in which the electron stream is provided by a hot filament. This class of tube calls for a much lower voltage. With the filament heated by a two-volt battery a satisfactory beam of high intensity can be produced with 1,500 volts in the second anode. The "cold" tube, on the other hand, requires from 50,000 to 100,000 volts. To regulate the intensity of the beam, the kinescope has a special control electrode or grid introduced between the filament and the first anode.

The incoming radio impulses from the transmitter cause a change in the normal electron flow. This disturbance corresponds exactly with variations in the modulated current at the sending station. The action is similar to that which takes place within the standard three-element radio tube when a varying voltage is impressed upon the control grid. When a positive charge is on the grid some of the electrons from the filament are attracted, thereby reducing the number of electrons that reach the plate. The grid of the cathode tube performs a similar duty in that it increases or diminishes the total number of electrons that strike the fluorescent screen, in accordance with the current variations received from the transmitter. Thus the image is seen at the receiver exactly as the original appeared at the television station.

THE KINESCOPE'S DESIGN.—Zworykin's kinescope is sealed in a cone-shaped bulb with a narrow neck. Part of the neck is silvered and so is the inner wall of the conical portion. A lead-in wire makes electrical contact with the silver coating. The slightly convex base of the cone, ranging from six to nine inches in diameter, is internally coated with a substance that makes it a fluorescent screen. The fluorescent film is a trifle conductive and makes electrical contact with the silvering to prevent an electric charge from collecting on the screen and thus repelling the electron beam or pencil. Therefore, the interior of the bulb is a completely enclosed conductive surface which acts as a second anode. It gives a final acceleration to the electron beam and at the same time focuses the beam into a small spot on the screen.

Focusing is accomplished by an interaction of the electrostatic field between the first and the second anode and the moving electrons. The focus is easily regulated by adjusting the ratio between the potentials of the first and the second anode. The focusing is not dependent on the presence of residual gas. The higher the vacuum the better is the focus. The point of concentration of the electron beam is moved closer or farther from the first anode by slightly changing the ratio between the first and second anode potential. The actual focus is obtained by bringing this point to coincide with the surface of the fluorescent screen.

The filament is of the indirectly heated type, which permits alternating current operation. Special precaution is taken in the construction to prevent the filament supply current's magnetic field from interfering with the electron beam.

The first anode is a part of the electron gun located in the neck of the bulb. It pulls the electrons away from the cathode (filament) and projects them into the conical section of the tube.

When applied for television purposes, the first anode potential is + 400 volts, the second anode potential + 2,000 volts. The control electrode is − 45 volts. The normal filament current is 1.6 ampere at 2 volts. By varying the voltage to the control electrode, the second anode current can be changed and consequently the strength of the spot on the fluorescent screen can be controlled. Since the controlling potential is small when compared with that of the second anode, the control of the intensity does not affect the deflection of the beam. This accounts for the successful use of the tube for reception of television pictures without distortion, even for strong contrasts of intensities.

The cathode ray tubes are not usually provided with inside deflection plates, but are operated by magnetic fields. The deflection fields are applied close to the first anode where the velocity of electrons is comparatively low. This makes the tube quite sensitive for deflection. However, when magnetic deflection is impossible, electrostatic deflection, by means of deflecting plates, is used.

Some refer to the television cathode ray tube as the Braun tube. This is not correct because it differs in many respects from the original tube as invented by Professor Braun. The main difference is in an additional controlling element which does not exist in the Braun tube, and which is necessary in television to modulate the intensity of the fluorescent spot. This addition makes as much difference between the original Braun bulb and the television receiving cathode ray tube as there is between the two-electrode Fleming valve and the deForest triode.

Two Schools of Thought.—It can be seen that two schools of thought in television are forming out of the various types of experimental work conducted during the past few years, the mechanical versus the electrical scanners. The disk or drum is the contributing factor in the mechanical method while the cathode ray tube is the heart of the electrical scanning system.

The main advantage claimed for electrical scanning is no moving parts, and, therefore, complete absence of noise. As opposed to this is the mechanical system, favored by some engineers because it works at lower voltages and affords definite control of all its elements. Before the advocates of this system will discard it in favor of electrical scanning, the cathode ray tube must be made to deliver a much stronger white light, instead of a comparatively feeble illumination of a greenish tint.

Hollis Baird asserts that the cathode tube requires about 2,500 volts for satisfactory operation, and even then the illumination is far from the intensity desired for projecting pictures on a large screen. In its present form it is an expensive tube, the cost being estimated at approximately $75.

TELEVISION IN NATURAL HUES

Its life is limited to about 100 hours. The pictures vary in size, some of them being about four by five inches. One of the problems in connection with the cathode tube is control of the brilliancy and shape of the scanning spot. Baird points to the fact that high voltage is needed for brightness but high voltage decreases the sensitivity. The filament control is critical.

Against the argument that the mechanical system has moving parts, the advocates of that method contend that there has been no objection to home talking picture machines because they have moving parts. And they are usually noisier in operation than the television disk. It is true that the scanning mechanisms have been large and cumbersome but progress is being made in developing smaller, more compact devices.

"One particular point about the mechanical method is that with improved scanning mechanisms and better light sources pictures up to two and three feet square are perfectly feasible for projection on a screen," said Baird. "The cathode ray is definitely limited in the size of its pictures by the prohibitive cost of developing a huge cathode bulb. Of course, both mechanical and electrical methods are of great interest and both are finding sincere adherents."

Baird, in further comment on a statement by a radio engineer that the electrical way was the obvious way and would be the one used, stated that the helicopter was considered by early inventors to be the logical way of flying but that until 1930 no flight with any sort of revolving wing machine had been successful. In the meantime success has attended the development of the fixed-wing type of aircraft. That such a parallel is possible in radio is his contention. He asserts that the mechanical system of scanning offers more possibilities for the experimenter because present results make it a logical contender for television honors.

Chapter Eight

FACES ON WIRES—FACES IN SPACE

An image has been flashed to Australia and back to the United States in the twinkling of an eye. Speakers at the ends of telephone wires are seeing each other as they converse. Images of men and women are dancing, singing and joking on a large theater screen upon which a television projector casts a beam of light.

It begins to look as if a new international theater is being built, a new industry created, a new link being forged in the chain of friendship between the nations of the earth. Soon man will see his fellows smile on the other side of the globe, while hatred and suspicion are torn from the imagination that functions when people cannot see what others are doing or talking about. Television knows no frontiers. It rips down barriers. It will empower man to shake hands across the sea, across the hemispheres.

It is 1930!

To Australia and Back in a Flash—February 18, 1930

The days are getting longer in the Mohawk Valley. Winter with its ideal atmosphere for radio is on the wane. An automobile carrying several research experts has left Schenectady and is on the way out through Scotia and up a long winding hill that leads to the top of an Adirondack foothill. In a little house at the summit vacuum tubes give a glow of warmth to this frosty morning. It is 8 o'clock. On the other side of the globe other vacuum bulbs are shining. Everything is ready!

An American asks an Australian if the waves are girdling the earth, and in a split second a voice with English accent answers with clarity that indicates the world-wide pathway is free of static. Short wave station W2XAF, Schenectady, has established communication with VK2ME at Sydney, 20,000 miles away. Dr. Alexanderson is on the job. He is ready to broadcast a television picture of rectangular design painted in black on a white card. He manipulates the electrical controls. There goes the picture and there it is back again before anyone can say "Jack Robinson." Schenectady projected it into space; Sydney picked it up and flashed it right back—all in one-eighth of a second!

Even the veteran engineers to whom radio magic is an everyday event marvel at the uncanny result and the terrific speed, which one might expect would rip a photograph asunder and scatter it through space.

"Considering the fact that this picture bounded through the air twice over so great a distance," said Alexanderson, "I am much enthused with the result. I really did not believe the picture would be distinct enough to recognize when it got back to us, because so many conditions lurked in its path to upset matters.

"There are ripples in the ether, such as there might be in a pail of water. When one looks into a pail of water that has been caused to ripple the reflected image is indistinct. The lines of the picture are exaggerated and made to appear fuzzy. In this rebroadcast, it was much the same as though this image seen in one pail of rippled water had been reflected in another pail of rippled water, corresponding to the rebroadcast back from Australia.

"Naturally there would be considerable distortion, and I was much pleased when I saw that this double distortion did not entirely wipe out the image. The test was carried on for about five minutes, and many times during that period the lines of the rectangle were distinct enough for observers to identify the picture being broadcast."

London Sees "Abbreviated Vaudeville"—April 6, 1930

When Sir Ambrose Fleming invented the two-element vacuum tube, back in 1904, probably he little realized that some day he would stand in front of a scientific machine designed to send the image of his countenance across the English countryside to be seen at least ten miles away. And the observers also hear him talk as he participates in a preliminary introduction to an "abbreviated vaudeville" performance being wafted across the housetops in the British Isles. Fleming steps away from the televisor. Gracie Fields, a songster, is ready to begin the vaudeville of the air.

One of the television receivers is installed at 10 Downing Street, the official residence of the British Premier, Ramsay MacDonald. Mr. MacDonald's daughter and other members of the household watch and listen to the performance. Twin broadcasters, operating on different wave lengths, are being utilized. One wave handles the image and the other the voice or music. The spectators see the head and shoulders of the person televised. The image appears in what is known as a television mirror.

The success of this demonstration causes Americans to wonder if Uncle Sam is being left behind in television. The radio leaders testify he is not backward; in fact, they say he is far ahead. Again they declare "television is in the research laboratories and is not likely to emerge for public use until it is commercially practical and foolproof."

The American engineers see numerous technical obstacles which must be overcome before images can be sent through space with the same clarity that makes listening to radio musicales a pleasure. It is not likely television will be introduced in the United States on an experimental basis as was broadcasting when thousands of receiving sets were built at

FACES ON WIRES—FACES IN SPACE   109

home. Radio is now an industry, and when television is ready to leave the research laboratories it will do so with greater perfection than did the early broadcast receivers. Factory built sets will be available to meet a nation-wide demand.

The images and sound in the ethereal vaudeville show were picked up ten miles away, according to reports from London. On the other hand when D. W. Griffith participated in a WGY transcontinental television broadcast, his image crossed the United States on the 21.96-meter wave, while his speech traveled on the 31.4-meter channel, as well as on the 380-meter wave of WGY. That enabled broadcast listeners to eavesdrop on his words. Only those with the proper short-wave television machine and short-wave receiver could see and hear him too. The program was on the air for fifteen minutes in February, 1929. Receivers in California picked up the picture and words that darted out from aerials in the Mohawk Valley.

JEWETT'S OPINION.—America is not being left behind by England in the matter of television development, according to Dr. Frank B. Jewett, vice president of the American Telephone and Telegraph Company. He is also president of the Bell Telephone Laboratories, Inc., research organization of the telephone company, which for a number of years has been engaged in investigations to determine if and how television might be adapted to modern life as an improvement in any existing commercial system. It is his belief that neither English nor American investigators have found a clue upon which to concentrate their endeavors.

"According to my belief," said Jewett, "television has not progressed beyond the experimental stage, and as for England leaving America behind, it is not so. For many years past we might have operated in this country a simultaneous sight and sound broadcast but in order to have a thing like television in the home one must show clear pictures, else after a while one would get tired of looking at them. No television at the present time is as good as the movies. As a result, one would always be comparing television with the motion pictures.

"First, it is difficult to operate television over any kind of a radio channel because of interference, such as static. A non-interfering vehicle is essential. Second, to make television a thing of enjoyment in the home today one must have elaborate and expensive apparatus. The matter of synchronizing the transmitter and receiver is no longer a problem. Television is a reality in America as it is apparently in London, but it is not commercially practical."

"This London television experience is certainly no novelty to American radio fans who for the past three months have been receiving television pictures synchronized with voice from the Jenkins television transmitter at Jersey City," said Lee de Forest. "In this case the voice has been broadcast on 187 meters and the television pictures on 142 meters. Thus far the nightly pictures transmitted have been those of talking motion picture film, but it will be a matter of only a few weeks before the visage and voice of visitors to the studio will be broadcast."

THE TIME HAS ARRIVED.—The Jenkins television laboratory at Jersey City reports to the Federal Radio Commission that the time has arrived for sight-sound broadcasts.

Lieutenant E. K. Jett of the engineering staff of the commission contends that experiments indicate television is still in the laboratory stage and any programs put on the air now would have little entertaining value, and would create only an "unrecognizable mess." However, it is pointed out that the early broadcasts of music were termed "a mess" by the opera stars and other noted artists who could not be persuaded to face the microphone. Tone quality in those days did not count. Radio was a novelty, just to hear a distorted voice or discordant music was heralded as wonderful and thousands rushed to buy radio sets or to build them.

The public will not be so particular about the clarity of the first television pictures.

Speakers on Telephone See Each Other—April 9, 1930

Each spring for the past few years has seen a new type of television blossom at the Bell Laboratories, where Dr. Ives is nurturing the seeds of radio vision. This season it is two-way television in which the speakers at both ends of a telephone line or radio circuit see the images of each other as they converse. The demonstration is conducted over wires between the American Telephone and Telegraph Company, 195 Broadway, and the Bell Telephone laboratories at 463 West Street, about two miles apart.

The system is applicable to radio but with less certainty than when wires link the two points. However, distance is no obstacle. The engineers point out that it is just as easy to let a person in San Francisco see a person at the other end of the line in Boston, New York or Philadelphia as it is to see the images over a shorter distance.

Special television booths have been developed which are about the same size as an ordinary telephone booth. Upon entering the booth the person to be "televised" sits in a swivel chair and faces a frame in which he will see the person at the other end of the line to whom he will speak. The face is illuminated by a mild glow of blue light reflected from the face to the photoelectric cells, known as the "radio eyes." This causes the current to flow and carry the image by wire to the distant booth.

There is no glare or flood of brilliant light as in early television systems. At first, as one enters the booth one notices a dim orange light which is too weak to affect the photoelectric cells. The usual telephone is missing. Special television transmitters and receivers are hidden from view. It was necessary to dispense with the ordinary phone because it would hide part of the speaker's face from the distant observer.

THE CURTAIN GOES UP.—When the speaker turns in the chair and faces the apparatus he sees on the glass screen the words, "Ikonophone—Watch this space for the television image." Then this sign lifts like a magic curtain and in its place the animated picture of the person at the other terminal appears. The two converse in ordinary tone as over the telephone. The images are about a foot square and are extremely clear.

This 1930 television image is greatly improved over that shown by the Bell Laboratories in 1927. It is double the size with more clarity and detail. The "radio eyes" are much more sensitive. They create ten times the amount of current for the same amount of light as did those of three years ago.

The dazzle of light has been eliminated by the increased sensitivity of the electric eyes and by the blue scanning beam. The person being televised never realizes that his face is being swept eighteen times each second by the beam of light that illuminates it. Both parties to the television-telephone conversation see each other with sufficient detail to recognize the facial expressions. It is like an instantaneous motion picture in black and white on a pinkish background caused by the color of the high-powered water-cooled neon tube utilized in the receiving set. No part of the system is annoying to the eye.

The voices are picked up by a sensitive condenser-microphone the same as used in broadcasting and sound-picture recording. The microphone and a small loudspeaker are concealed behind the screen upon which the image appears. The microphone is located a trifle above the head and the loudspeaker about even with the knees of the person in the booth. Both are invisible to the persons using them.

COMMERCIAL ASPECT UNCERTAIN.—"Despite the fact that the research and development work of the past three

FACES ON WIRES—FACES IN SPACE 113

years has resulted in a great improvement and simplification of the equipment required for television," said Walter S. Gifford, President of the American Telephone and Telegraph Company, "it is still necessarily complicated and expensive, requiring expert attention and large units of apparatus. These facts arise out of the inherent technical requirements for satisfactory television transmission. While substantial progress has been made on the technical side, the future commercial possibilities of television are still uncertain. In line with our long established policy of fully exploring and developing every field which gives promise of possible improvement in extension of electrical communication we expect to continue our television work."

LIGHT IS DIFFUSED.—Scanning is performed by the beam method. The scanning beam is derived from an arc lamp the light of which passes through a disk that has a spiral of holes. Then the light beam passes through a lens on the level of the eyes of the person being scanned. The light reflected from the face is picked up by the array of photoelectric cells which are in the television booth behind plates of diffusing glass. The current from the cells is amplified and sent by wire to the receiving station.

The received signals are translated into an image by means of a neon glow lamp directly behind a second disk driven by a second motor placed before the first motor and disk used in the transmission. The two disks are inclined at a slight angle to each other. The disks vary in size. The upper one used for transmission is twenty-one inches in diameter. The receiving disk is thirty inches in diameter. The disks used in earlier demonstrations had fifty spirally arranged holes. Some of the later disks have seventy-two holes so that the image detail is doubled, in fact, there is never any doubt about recognizability. Individual traits and facial expressions are unmistakably transmitted.

MAKING THE PICTORIAL CALL.—From the standpoint of the user, the engineers have succeeded in simplifying the operation of the combined telephone and television. A person enters the booth, closes the door, sits in a revolving chair, swings around to face a frame through which the scanning beam reaches his face, and upon seeing the person at the other end of the line, he talks in a general tone of voice, and he hears the image speak. The conversation is carried on as if the two people were at opposite sides of a table.

"Some of the more special problems encountered in two-way television are primarily optical in character," said Dr. Frank Gray. "The principal one is that of regulating the intensity of the scanning light and of the image which is viewed so that the eyes are not annoyed by the scanning beam. And precautions must be taken so that the neon lamp image is not rendered difficult of observation. It has been necessary in the solution of this problem to reduce the visible intensity of the scanning beam considerably below the value formerly used and to increase the brightness of the neon lamp.

"The means adopted consists, first, in the use of a scanning light of a color to which the eye is relatively insensitive but to which photoelectric cells can be made highly sensitive. Blue light is used for this purpose. It is obtained by interposing a blue filter in the path of the arc light beam. Potassium photoelectric cells specially sensitized to blue light and more sensitive generally than those previously employed have been developed. The number of these cells and their area has also been increased over those utilized in the earlier television apparatus. Thus the necessary intensity of the scanning beam is decreased."

The second half of the problem—namely, that of securing a maximum intensity of the neon lamp—has been attained by the development of water-cooled lamps capable of carrying high current. The net result of the blue light for scanning, the use of more sensitive photoelectric cells, and the high efficiency neon lamps is that the person being televised is subjected only to a relatively mild blue light sweeping across the face, which he perceives merely as a blue spot of light above the incoming image.

OBTAINING PROPER ILLUMINATION.—A second optical problem is the arrangement of the photoelectric cells in order to obtain the proper illumination of the observer's face. The photoelectric cells act as virtual light sources. They can be manipulated as to both size and position like the lights employed by a portrait photographer in illuminating the face. In the television booth, it is desired to have the entire countenance illuminated and, therefore, photoelectric cells are provided at either side and above the person in the booth.

One practical difficulty encountered is that eyeglasses, which often cause annoying reflections in photography, act the same way in television. It is important, for this reason, that the photoelectric cells be placed as far to either side or above as possible. Then the reflections from eyeglasses are not annoying unless the person turns his face considerably to one side or the other. The number of cells has been so chosen as to secure a good balance of effective illumination from the three sides. It has been found desirable partly to cover the cells on one side of the booth in order to aid in the modeling of the face by the creation of lights and shadows in one direction.

Illumination of the interior of the booth presents another optical puzzle. There must be sufficient light for the user to locate himself. It is also desirable that the incoming image and scanning spot is not seen against an absolutely black background. The booth is illuminated by orange light to which the photoelectric cells or "eyes" are practically insensitive. The walls and floor of the booth are well illuminated. A small light is provided on the shelf bar in front of the observer so an orange light is cast on the front wall surrounding the frame in which the picture appears. This light contributes materially in reducing the glaring effect of the scanning beam, and facilitates visibility of the incoming image.

LARGE CELLS ARE SENSITIVE.—Each photoelectric cell is twenty inches long and four inches in diameter, giving it an area of approximately eighty square inches for collecting light. The sensitive cathode consists of a coating of potassium sensitized with sulphur, covering the rear wall of the tube. This type of cell is more sensitive than the older "eyes" that utilized potassium hydride. To amplify the photoelectric current, the cells are filled with argon at a low pressure. Electrons passing from the sensitive film of potassium to the anode ionize the gas atoms along their paths and thereby cause a greater flow of current.

Twelve large photoelectric cells are mounted in the walls of the booth. They present an area of approximately seven square feet to collect light reflected from the subject being televised. A group of five cells is located in each side of the booth. Two cells are in the sloping front wall above the person in the booth. All cells are enclosed in a large sheet copper box, provided with doors to each group.

An operator is on duty behind the compartment to insure that the incoming and outgoing images are properly positioned, no matter what the stature of the person sitting in the booth. He must adjust the images to proper clarity. The optical monitor adjusts the scanning beam and position of the viewing lens to suit the height of the sitter.

TIGHT-ROPE ROUTE IS SAFEST.—Television images at their present age should be content to walk the straight and narrow path on the wire line rather than to take a long run on an invisible radio wave. Ives warns them to stick to the wires until engineers can entrust them to the more uncertain radio channels. Man has no control of the images once they

FACES ON WIRES—FACES IN SPACE

enter the portals of the ethereal realm, but when they travel by wire he can do more to direct their destiny.

To send television images through space today is like expecting a brook running through a populous area to remain uncontaminated, according to Ives. But he is hopeful. He has faith in man's inventive genius. He foresees the day when the limitations of radio will be overcome. But until then the images are safest when they stay on the wires. Then they are out of range of nature's shots. The minute they leap from a radio aerial they are at the mercy of elements out of man's control.

"Two-way television requires the equivalent in wire lines which would carry thirty ordinary telephone conversations," said Ives. "To accomplish the same by radio would require at least fifteen to twenty wave lengths. Wires may be crisscrossed through cities and be kept comparatively free of interference. It is an entirely different proposition with radio. In television broadcasts the images may encounter all kinds of interference. Networks of wires may be utilized for television and, with a little care, be kept clear of outside influences that might mar the images. Natural interference cannot be entirely averted, in the present state of our knowledge, unless the entire channel is definitely under our control at all times. A wire is the only thing which we can bring under this classification."

LIKE SLICES OF BREAD.—Ives says that radio waves, apart from their susceptibility to natural sources of interference, must be shielded from each other by separation in the radio wave spectrum. Some sort of a "fence" must be erected between the waves to stop any interference that one broadcast might cause by mingling with another. This "fence" is nothing more than separation of the waves. In other words, the radio wave band is not separated like a loaf of bread after a knife divides it into slices. The separation between television waves must be like removing every other slice in the loaf. But the radio "loaf" cannot be stretched out. This is one reason why it is impossible at the present time adequately to put television on the air from a great number of stations. The waves available are scarce. Ultra-short waves hold promise of solving this problem.

Television Goes on the Stage—May 22, 1930

Television images are performing on a theater screen in a world première. They dance, sing and joke.

It is a great day for Dr. Alexanderson, who for years has been giving these ethereal actors the proper electrical nourishment in his laboratory at Schenectady, so that they might grow from dwarfs to the life-size of real Hollywood stars. They wink and blink, as if bidding for a welcome into the American home. Their appearance on the big screen reveals that the wizards at "the House of Magic" have realized their ambition to build up the images from the size of the face on a dollar bill to the natural size of man. It was last autumn that faces were shown on a screen fourteen inches square, but now the screen measures six by seven feet!

Is It Only a Dream?—Often, Alexanderson is asked if television will ever be practical, or if it is only a dream. He always smiles and shrugs his shoulders as he answers, "Oh, television is a long way off, three years, possibly five or ten."

But now, in 1930, Alexanderson like a magician waves aside the veils of secrecy and shows, on the stage of Proctor's Theatre in Schenectady, television performing tricks that astound the audience. Vaudeville teams banter back and forth. One member performs and jokes before the televisor, while the other replies from the stage. Duets are sung by vocalists two miles apart. The theater orchestra in the pit is directed by a conductor who waves his baton on the screen. He is two miles distant. Local newspapers advertise the first television show ever staged in a regular theater. And a capacity audience attends.

Courtesy Bell Telephone Laboratories

MAN'S ATTEMPT TO COPY THE EYE

The large square comprised of 2,500 tiny cells might be likened unto the retina of the human eye while the 2,500 wires leading back from it represent the optic nerve in the television system. This was the track the engineers were on in 1927 when Herbert Hoover was seen in New York by television from Washington.

FACES ON WIRES—FACES IN SPACE

The lights are dimmed as in any motion picture playhouse. The curtains part. In the center of the stage is a screen. At the side stands a man with a telephone. He calls the television studio two miles away, and the audience hears him announce that all is ready for the performance to begin. The telephone is utilized to convince the audience that it is a real television performance, and not a talking picture on a film. A stage manager uses the telephone to direct the distant actors. A flood of light washes across the screen. It wavers and flickers like the early movies. A face appears. It is Merrill Trainer, who is acting as master of ceremonies. The picture is clear. There is a thunder of applause from the audience. Trainer hears the cheers through the telephone in the hands of the man at the footlights in the theater. He bows, smiles and thanks the audience for the enthusiastic greeting.

RECRUITS FROM VAUDEVILLE.—The stage manager asks Trainer to smoke a cigarette. He takes a pack from his pocket, scratches a match and blows smoke rings across the screen. Then the entertainers take their turn at the televisor.

Matilda Russ, a soprano, flashes on the screen. The voice reproduction and accompaniment are excellent. Entertainers recruited from the vaudeville circuit next parade in front of the televisor's eye. Two who usually appear in blackface do not blacken up for this performance, because television to-day does not take a blackface, although it may later. If a television actor wants to appear on the screen in blackface he must use green paint instead of burnt cork.

The television screen is wheeled out on the stage just as easily as a piano for a novelty act. No longer is the image restricted to the miniature dimensions of a postage stamp or a postal card. No longer must the observer squint through a tiny peek hole to catch a glimpse of the fleeting scenes and characters. And behind or at the side of the screen is television's voice, a giant loudspeaker which reproduces the voice of the speaker several miles away, at the same time that his actions and facial expressions appear on the screen.

AN INNOVATION IN PROJECTION.—Unlike the movies, no beam of light streaks across the auditorium above the heads of the audience. Television's projector is located backstage. The pictures are thrown on the screen from behind. This new art seems destined to introduce innovations in theatrical entertainment.

The actors are televised in an improvised studio which is a part of the laboratory. The light reflected from the faces is converted into electricity and then into radio energy broadcast by a laboratory transmitter tuned to release the images on the 140-meter wave. Microphones close by pick up the speech, music and songs, and convert the sound into electricity, which is carried by wire to a short-wave transmitter at South Schenectady for broadcasting on the 92-meter wave. The sound and images are scattered through space at the speed of sunlight. Over at the theater a control operator is busy. His duty is to manipulate the apparatus that intercepts the moving pictures that are somewhere in the air. A small device called a monitor telopticon transfers the impulses to a light valve, at which point the light is broken up to produce the image that corresponds in every detail to the person or object being televised several miles away.

THE KAROLUS LIGHT VALVE.—The light valve is based upon the invention of Dr. August Karolus of Leipzig, Germany. It is the heart of an intricate system of lenses, which is in front of a high intensity arc lamp similar to those used for the projection of motion pictures. The light valve is a delicate device. It is used instead of a neon tube. It must function with the utmost accuracy to permit the passage of light that corresponds perfectly with the impulses received from the television transmitter. These light emissions are passed on through lenses to a disk corresponding in size, design and rate of rotation to a disk at the radio "camera" or originating point. Other lenses pass the light forward to the screen, on which the light impulses, at a rate of 40,000 per second, wash or paint an active, life-like motion picture.

Karolus has modified and improved the Kerr cell. Kerr, an English physicist, discovered the principle that certain insulating materials or dielectrics rotate the plane of polarization of a light ray between two prisms when subjected to electric strain.

Nitrobenzene, carbon bisulphide and other substances are highly refractive dielectrics. If two metal plates are suspended in their medium, the plane of polarization of light passing between the plates can be rotated by sending an electric potential across the two plates. If such a device is inserted between a pair of Nicol prisms, it becomes an effective light valve. The light is then modulated in accordance with the applied voltage. The Karolus valve employs nitrobenzene, which, incidentally, is a poison that can be absorbed by contact. Despite the advance that this valve makes possible the engineers want a device that will pass a much more powerful light. If they can get that they can enlarge the pictures without sacrificing clarity.

IMAGES ENTER A TUNNEL.—The arc lamp, with the associated lenses and light valve, which all comprise the television projector, is placed seventeen feet back from the screen. A heavy black cloth from the projector to the screen forms an effective light tunnel or hood, which eliminates stray light beams that might blot or blur the pictures. The entire apparatus is mounted on wheels to facilitate assembly and disassembly when used as part of a vaudeville show.

A second receiver on duty at the theater detects the words or music, which are fed into the large loudspeaker—television's voice.

The life-like image is not a silhouette, nor is it merely a black and white picture. All the gray shades between black and white are reproduced, registering every shadow and shade of the original scene, giving both depth and detail to the image.

It is well to remember that in radio broadcasting the frequencies of speech and music modulate or shape the current sent out from the aerial wires. In television the aerial radiation is modulated or formed to correspond to the image by a succession of light impulses. The person to be televised stands in front of an incandescent lamp. Between the person and the light is Nipkow's metal disk about the size of a bicycle wheel and drilled with forty-eight holes. The disk revolves so that it covers the person's face twenty times in a second. That creates twenty complete pictures made up of light and shade. A large square frame contains four photo-electric cells. These "eyes" respond 40,000 times in a second to the light impulses reflected from the person being televised.

FORCES THAT CREATE EPOCHS.—"Looking back over the development of the electrical industry," said Alexanderson, "we can clearly trace the forces which have enabled the science of electricity to give birth to the electrical industry. We see how later the electrical industry took hold of another branch of science and created the radio industry. We are able to some extent to project into the future the working of the forces that give birth to new epochs, but as to the destiny and significance of these new movements, after they have been launched, the engineer is peculiarly blind. Owen D. Young has repeatedly said that he has the great advantage of not being handicapped by scientific knowledge. His predictions of the future have been much more far-flung and correct than those of the engineers associated with him.

"For fifteen years radio was simply an auxiliary to navigation. In 1915 and 1916 we held daily communication by radio telephone from Schenectady to New York. We found that many amateurs adopted the habit of listening, and our noon hour of radio became the first regular broadcasting.

But we had no idea to what it would lead. Our idea was to telephone across the ocean, and so we did at the close of the war, but we failed to see the great social significance of broadcasting.

"Television is today in the same state as radio telephony was in 1915. We may derive some comfort from this experience of the past, but, on the other hand, we are not sure that the analogy is justifiable and that television will repeat the history of radio telephony. We must then fall back upon our conviction that the development of television is inevitable on account of the forces working in the scientific world today, and that it is a satisfaction to make one's contribution to this evolution even if, in this case, the results should prove to be only a stepping-stone to something else."

WHO INVENTED TELEVISION?—Alexanderson is asked to name the inventor of television. He replies that the nearest to a simple answer is that Nipkow invented television more than forty years ago. However, Nipkow lacked the radio amplifier, neon lamp and photoelectric cell. Therefore, his invention could not be completed at that time. It remained for others to overcome numerous obstacles. Nipkow did, however, clearly explain the idea of scanning the picture, line after line, by a spot of light.

"Before we could produce these 1930 results," said Alexanderson, "we had to make several tests with different wave lengths. Many of them proved to be failures because one ray or wave followed the surface of the earth, whereas the other was reflected from a layer of electrons 100 miles above the earth. We are now working with the 140-meter wave, in which the ground wave is predominant. On the other hand, for long distance, we have found it advantageous to use the shortest possible wave lengths, so that the bulk of the radiation leaves the earth and only the lower fringe of it will arrive at the receiving station. It is expected that the tests, now in progress, will throw more light on the subject of wave propagation.

"Television apparatus is an ideal working tool for experimenting, and I venture to predict that we will soon see a wave of activity in amateur television. There are more than 100,000 experimenters in America, young and old, who go in for radio not to be entertained but to build their own sets and get a thrill from exploring the unknown. These amateurs have been rather starved of real interest in the last few years because of the commercialization of broadcasting. They will popularize long-distance television just as they created an early interest in broadcasting.

INCENTIVE FOR EXPERIMENTERS.—"The amateurs and the professional experimenters are on common ground. We got a real thrill out of sending a television wave to Australia and have it come back and tell its tale, even though it was a simple one. We observed that after traveling 20,000 miles a rectangle still had four corners, which was more than we had expected. As a matter of fact, it was broken up into pieces most of the time. But there were glimpses of encouragement and a fertile field for the imagination. These are the incentives of the explorer, whether he is an amateur or a professional.

"Whether the general public will be enough interested or get enough satisfaction out of television to make it possible to commercialize home sets is still to be seen. A new technique of entertainment will be required. As a supplement of broadcasting it can make a reality of radio drama. Political and educational speakers may use it as a medium, and entertaining personalities like Will Rogers will tell the latest wisecracks and comment on the news of the day. It is likely that every moving-picture theater in the large cities will be equipped to give a short television act.

AN INTERESTING RACE.—"What we have demonstrated is just one of the many steps that must be taken in our efforts to conquer distance by television. The improvement of light control which makes it possible for us to show a picture of theater size is due to the light-valve invention by Dr. Karolus, whom I visited in Leipzig some years ago and whose inventions we have been endeavoring to perfect. In our past exhibits the improvements of light control have been due to Dr. D. McFarland Moore and his neon lamps.

FLYING NEWS REPORTERS.—"The possibilities for new inventions in television are inspiring," continued Alexanderson. "Just think of what can be done when you can put an electric eye wherever you wish and see through this eye just as if you were there. An airplane with a news reporter will fly to see whatever is of interest and the whole theater audience will be with him, seeing what he does, and yet the audience will be perfectly safe and comfortable.

"What will this mean in the wars of the future when a staff officer can see the enemy through the television eyes of his scouting planes or when a bombing plane is sent up without a man on board to see the target, drop the bomb and be steered by radio? What will it mean for peaceful aviation when the ships of the air approach a harbor in fog, take on a local pilot, not from a little craft that comes to meet the ship, but by television, whereby the trained eyes of the pilot functioning by television will guide the ship to the airport in safety?"

Alexanderson does not expect that seeing by radio will give as much detail as a talking picture. Television gives immediate action and is not what he terms a "canned" show. He believes, however, that television will eventually picture football games and news events when a radio camera is on the scene.

"Television will be a great asset to politicians," he said. "However, they will have to prearrange their speeches to conform with broadcasting schedules. The day is likely to come when candidates for President of the United States will campaign by television. The winner may be elected because of a winning smile that enters the homes of millions. I do not want to predict when we will have television in the home. All I can say is that we are continually making good progress."

CHAPTER NINE

A CASTLE AND A CITY OF DREAMS

There are fortresses on hilltops and cliffs throughout the Old World, and fortified strongholds in the Land of Dreams, but none so electrified with modern ideas as a magnificent castle the turrets and spires of which project from thick stone walls on a rocky headland along the seashore at Gloucester, Mass. That is the scientific mansion of John Hays Hammond, Jr. In that medieval castle he is busy solving modern problems of radio, television, music and aviation.

And on Manhattan Island a tremendous hole is blasted in the rock into which tons and tons of concrete are poured, above which steel fabrics are woven. The riveters peck away like woodpeckers while masons follow them skyward to cover the steel cage and framework with millions of bricks, tiles and stones. And in the end this ornate structure will have an urban landscape, hanging gardens, on the grandest scale ever attempted since the days of Babylon.

Television is an inspiration to art and science, to financiers and builders, to showmen and to artistry!

AN INVENTOR'S GLIMPSE OF THE FUTURE—JUNE 10, 1930

John Hays Hammond, Jr., is a pioneer in radio dynamics. And now he has invented a television eye for airplanes so the pilot can "see" the landing field and surrounding terrain no matter how thick the fog or how dark the night. He is developing some radically new ideas that seem destined to improve the sound reproduction of phonographs, pianos and talking pictures as well as television. He is a man with an international reputation won by his radio controlled boats, vehicles and torpedoes.

In his castle by the sea John Hays Hammond, Jr., dreams dreams that come true. Looking out over the ocean through the narrow, slit-like windows of his laboratory, this radio inventor meditates and plans for new scientific wonders to benefit mankind. He has discovered that the most fruitful ideas from which big strides in progress evolve are simple. They flash upon the mind in odd and unsuspecting moments.

The visitor who calls at this unique workshop of science first must cross the wooden bridge that spans the moat before he can rap on the big iron door, an embattled gateway that guards the inner secrets. The main room of the castle is of large proportions. It is like a great Gothic church with all the pews removed. There in a little chapel at one side of the spacious room the inventor greets his guests.

One might expect to meet a bearded scientist garbed as an alchemist of yore. But Hammond looks more like the leader of the Yale Band, in his coat of New Haven blue, a dark blue tie, a white shirt and white trousers. 'Tis true the laboratory has an ancient setting, but the inventor is modern. His numerous problems and ideas are ultra-modern. He is always looking ahead.

NEW WONDERS FORESEEN.—"We ought to have a thousand research workers here instead of a few as we have," said Hammond, "because we have so many ideas to be developed. The span of life is short and affords us opportunity to get only a start for what the next generation will achieve. Radio is just beginning. And so is television—although I applied for a patent on color television fifteen years ago, only to find later that some one had beaten me to it by more than ten years.

"Radio vision is here today, if we do not attempt to span too long a distance. I believe that before television goes into the homes it will be seen in theaters and auditoriums in the

A CASTLE AND A CITY OF DREAMS 129

large centers of population. For example, there is one popular theatrical performance in New York at which many have been unable to get a seat. In connection with such a popular stage production, why not rent three or four other theaters along Broadway and in them produce the original play on a television screen? It would be almost as good as the original. The box office could charge a little more to see the original than the duplicate. But, in the end, more money would be made because more people would have an opportunity to see the performance.

"The Yale Bowl, Harvard Stadium, Yankee Stadium, Polo Grounds and Palmer Stadium at Princeton can hold just so many. Thousands are turned away from the big games. And thousands of enthusiasts in cities miles away cannot attend in person. So, I foresee television bringing the major sports events in the East to capacity audiences watching the contests on television screens in Detroit, Chicago, San Francisco, Boston and other large cities. Then, the next step will probably be into the home. However, to be practical and economical the television impresarios ought to have a pay-as-you-enter plan before they go on the air."

It will be recalled that soon after the broadcasting "craze" swept the country Hammond suggested a method to make programs available only to those who had the right tuning "key," and he told the infant radio industry how it could become a big business on an economical and self-sustaining basis. The leaders of the radio industry, however, objected to broadcasting being operated on a toll principle. It was not long before the broadcasters realized that Hammond was right, for all of them were losing money. Some dropped by the wayside, then the advertisers came to the rescue and bought time on the air.

GIVING AIRCRAFT EYES.—"Today I am devoting much of my time to a television application that safeguards aircraft landing in fog or darkness. It is such a simple idea," he said smiling, "and these are always the last to hit upon."

Three radio compass stations are located alongside the airport or flying field. The plane carries an automatic radio transmitter which sends out a continuous signal. Operators at the compass stations train the radio direction finders on the plane. The bearings are automatically recorded and sent by wire line to a television station near the field.

At this station is a miniature of the field, perfect in every detail. It shows every hill, tree, hangar building, fence and wire, exactly as they are laid out near the field. This model map corresponds to the surrounding terrain and over it are three movable arms. These arms represent the directive lines of the radio bearings.

Where they intersect is a television eye. That eye is in the exact position over the model field as the plane is above the ground. Therefore, what the television eye sees is the same as the aviator would see if his sight could penetrate the fog or darkness. The miniature field is, of course, indoors, so no weather can ever affect it. As the plane moves the arms move and the eye roves accordingly across the replica of the airport.

WHAT THE AVIATOR SEES.—The miniature scene that the eye beholds is televised and flashed to the plane's pilot. He sees everything on a television screen located ahead of him on the instrument board. As the plane moves, the pilot sees before him the exact scene that would be before his eyes if his vision were clear. He is, however, looking at a model instead of reality. Every detail of the ground below him is faithfully reproduced, every shadow and angle, with a reality that a well-executed model can give. He sees approaching a spire, or clearly defined telegraph wires—even the position of other stationary planes on the field can be added to the model and transmitted aloft.

As the plane turns, the pilot watches the change of scene

A CASTLE AND A CITY OF DREAMS 131

before him. His movements are continuously followed, so the scene before him changes continuously. It is immaterial to the pilot whether he scans reality or a perfect copy of reality. He is interested only in his position relative to the earth and the objects upon it, and this he constantly sees clearly defined as a view, his own bird's-eye view of everything below him.

It is by these combinations of well-known and tried principles that a new method presents itself by which a substitute vision is given pilots and the presence of fog, smoke or darkness becomes no longer a menace to life. These same principles can apply to shipping in the entrance to harbors.

Incidentally, an arrow at the center of the flying field as it appears on the screen reveals the direction of the wind. A number at one end of the field reproduced on the screen indicates the wind velocity, while another number at the opposite end of the screen indicates the plane's altitude. All other navigational instruments will remain in the televised plane because such equipment will be necessary in all territory where the televisor system is lacking.

HIGH POWER ESSENTIAL.—The main questions dealing with television progress as seen by Hammond are: How distinct is the image reproduction? How far can a scene be broadcast? He answers them by asserting that high power television solves both problems. It covers several miles with a clear image and overcomes fading.

"There is absolutely no doubt that television is applicable today over short distances, and by that I mean up to about five miles," Hammond said. "Alexanderson has developed the television technique which makes it easy to equip airplanes with all-seeing radio eyes."

Television requires from eight to ten times as much "space" in the air as radio broadcasting. For example, if the highway of music in space is ten feet wide, the road over which the images travel must be about 100 feet. Space in the ether is limited. Every available wave in the broadcast band is occupied. It is no wonder that the engineers are puzzled where room will be found for television. It is not only a question of developing television apparatus but how to make a path along which the images can dance through space.

Radio must be relieved of congestion. Hammond is interested in that problem. Already he has sent eight wireless messages on one wave length. He has demonstrated this from short wave lengths up to 1,700 meters. However, the degree of packing varies with different wave lengths; that is, there is more room for messages in the short-wave realm. They can be packed closely. The broadcasters are now overcrowded between 200 and 550 meters. What the Hammond invention does in this situation is described as similar to opening a 1,000-room annex to a 50-room hotel. The invention is compared in economic aspects with the discovery of the multiplexing system in telegraph, telephone and cable work.

This young inventor, who has also developed a method whereby aircraft may project torpedoes and then control their path in the water by radio from a high perch in the sky, asserts that the great problem facing all branches of radio is perfection of the fundamentals, and especially in transmission.

THE EYE IS CRITICAL.—"Time will come when static and fading will be conquered," he continued. "Broadcasting will be conducted with the efficiency of transatlantic telephony and a greater number of stations will fill the air. Within the next five years I think we will see television in theaters and auditoriums to take care of overflow audiences at national events. I mean over short distances.

"We must remember the eye is more particular than the ear. A crash of static now and then does not bother the ear so much, but let static freckle a television picture and the

A CASTLE AND A CITY OF DREAMS 133

eye will become mighty critical. It is going to be a bigger job to please the eye than the broadcasters have had in catering to the ear. The majority are eye-minded."

'Way down in the short-wave spectrum, where wave lengths are measured in inches rather than feet or meters, the farsighted scientist of Gloucester visualizes vast possibilities. The ultra-short waves are an unexplored region in radio. It is a new field that is calling for investigation. What scientists will find in this ethereal field, he hesitates to predict. They may find how to transmit more efficiently with less power and with simple apparatus. They may learn much about the construction of the atmosphere. They may discover new radio aids to medicine. Already experimenters have found that artificial fever can be created by short radio waves. Most microbes can live only in certain temperatures. Ultra-short waves may be the source of a fever that will kill certain germs without harm to the patient.

A LIFETIME STUDY.—"There is a wonderful future for radio," Hammond said. "It is far more than a lifetime study, we have so much to learn. Now that we have succeeded in promulgating the wedding of two great sciences, radio-television and aviation, we have indeed taken an important step. But radio research is just getting under way. Every step leads to new scientific applications.

"What a great thing if during the World War an airplane, through a television eye, 10,000 feet up in the sky, could have photographed the scene of a battle fleet even over a 100-mile area and then flashed that picture by television to submarines lurking below the surface! We can do that now. Such television maps in future wars will make it unnecessary for the tell-tale periscope to bob up as a target for enemy ships. Television will carry the surface scene far below the waves of the sea. The submarine maneuvers will be directed by planes far overhead and out of range of anti-aircraft guns carried by battleships. The winner of the next war will win because he has radio and aviation on his side."

Hammond, now in the early forties, is called "a chip off the old block." He is a graduate of Sheffield Scientific School at Yale. At the age of 34 he had more than 250 inventions to his credit, and his patents total more than six hundred. His father, also a Yale man, is known internationally as a mining engineer who built up a vast fortune from mining and engineering projects. Like father, like son, this radio inventor is prospecting in space as his father did before him in the depths of the earth.

A New City Rises on Manhattan Island—June 22, 1930

There is usually a reason for fabulous cities aside from their geography. New York has its harbor with all the world at its door by rail or ship. Albany has the Hudson River and the Erie Canal. Niagara has the falls. Chicago has its lake port, stockyards and railroad terminals. San Francisco's golden gate welcomes the Pacific and the great Far East. New Orleans has the mouth of the Mississippi and the Gulf of Mexico spread out before it. Minneapolis has the grain fields. And Kansas City has the railroads, wheat and cattle to make it a busy place.

Ever since America was carved out of the wilderness its thriving cities, towns and villages have sprung up and flourished because of water power, railroads, ports, wheat, cattle, fur, grain, lumber, gold, fruit, climate, quarries, fish, scenic beauty and what not. Now, because of invisible vibrations in the air, a city within a city is growing on the island that the Indians sold to Peter Minuet for twenty-four dollars. The new metropolis is to cost $250,000,000.

Foreseeing the dawn of a new era in electrical entertainment and education, and looking ahead to television with its vast possibilities, John D. Rockefeller, Jr., and a group led by the Radio Corporation of America, designed Radio

City or Rockefeller Center, to cover three city blocks in the heart of New York. It is bounded by Forty-eighth and Fifty-first streets and by Sixth and Fifth avenues.

"The sociologist's conception of a city has been a municipal unit, self-contained, with a more or less definite trading area, spreading its economic influence over as much of the surrounding country as can be conveniently reached by newspapers, railroads and motor cars within a few hours," remarked Dr. Alfred N. Goldsmith, when the plans for this magic community were announced.

"Now comes a city sired by science, mothered by art, dedicated to enlightenment and entertainment. It exists not for an immediate trade territory but for the world. Its drama and its dreams will be flung across oceans and continents. It will share its conceptions of beauty and culture with the farmer, the village store and the schoolrooms as well as with aristocratic foyers."

Radio, as it stepped from the dots and dashes of wireless to the voice of broadcasting, created a new art that won instant public acceptance. At first it was called a novelty, a luxury of entertainment. It amused. Voices and music that seemed to come from nowhere into the home with entertainment, education, religious services, news and music captured the imagination of the people. Today radio is called a household utility. Listening-in is part of home life.

IN THE BEGINNING.—Little did the KDKA pioneers in November, 1920, realize, when they broadcast the first program from an amateur station in Dr. Frank Conrad's garage in Pittsburgh, that radio was destined to grow into a vast industry; into a center of entertainment, which every American could enter by merely a snap of a switch and the turn of a dial.

Radio has advanced step by step. Each time it has moved, the new studios have been lauded as the broadcasters' Utopia, the best that the science of radio and acoustic engineering could offer. But the rapid pace of science has always pushed the broadcasters into new realms almost before they could get established in the old. For example, WJZ in 1921 began its career in humble quarters in Newark, N. J. Later it was removed to New York into elaborate studios in Aeolian Hall, then on Forty-second Street. Surely, it was believed, WJZ had found its ultimate home. But a few years later the station moved again, this time to 711 Fifth Avenue, to share with WEAF, the new headquarters, heralded as designed to accommodate radio for many years to come. But the restless radio nucleus was soon to make another move.

Ten years ago broadcasting was just getting under way. Wherever a transmitter went on the air there sprang up a demand for receiving sets. Existing wireless manufactures were not equipped to supply the demand. Thousands of amateurs built receivers on their work benches in the cellar, in the attic and on the kitchen table.

The theaters were warned to fight the menace. So were the motion picture and phonograph industries. The theatrical people continued to call radio a craze, a novelty that would soon wear off as did mah jong. It was natural that there should be skeptics. The theater was an established institution. It would beat radio when the novelty wore off. Producers said they were not afraid of this invisible competitor. Nevertheless, some saw the handwriting on the wall and were quick to link themselves with the new enterprise.

THE FRIEND OF ALL.—Today radio is a friend of all. It is bringing them all together in a city within a city. It has proved itself an ally of the theater. Its electrical devices have assisted in the development of the sound motion picture. It has electrified the phonograph and has given it renewed life, superior tone and new possibilities as a musical instrument.

The radio pioneers looked ahead. They planned for the future, just as they are planning today for greater triumphs in years ahead. Broadcasting is an art and an industry. This

A CASTLE AND A CITY OF DREAMS

is shown in the fact that the American public in 1929 spent approximately $850,000,000 for radio instruments. It was less in 1930 and 1931 because of the business depression.

There are more than 600 broadcasting stations in the United States. Many of them are linked by land wires for simultaneous broadcasting of the same program from coast to coast and border to border. There are more than eighty transmitters in the regular hook-ups of the National Broadcasting Company and more than eighty in the Columbia Broadcasting System. It is estimated that the waves from these stations reach every antenna in the Union. On special occasions, such as an important address by the President of the United States or a national political convention, the networks are expanded to take in other stations. Some are short-wave transmitters that send the events to foreign lands.

In 1922, when the theaters were beginning to wonder how they might "take over" radio and maintain control of it, because listening-in was showing signs of being more than a craze, there were about 60,000 receiving sets. Today it is estimated that the number is close to 12,000,000. It is believed that the average number of listeners per set is three, but the audience swells to much larger proportions when a heavyweight championship bout or some event of national interest goes on the air.

It may run up to 50,000,000, although no one knows.

MILLIONS OF DOLLARS INVOLVED.—Broadcasting is a business. The National Broadcasting Company for 1929 reported a gross income of $15,000,000; $22,000,000 in 1930. The investment in broadcasting runs into many millions of dollars. For example, the cost of a transmitting installation such as that used by KDKA, WLW and KMOX is estimated at $500,000. More than thirty-two thousand miles of telephone lines link the networks into a nation-wide chain.

Compared with this, in 1921, WJZ was housed in a small

building erected on the roof of a factory building. It was called an experimental transmitter, and used 500 watts of power, which in those days was considered high. Today the big stations use 50,000 watts, and WGY at Schenectady, which has tried 200,000 watts, has plans to experiment with 500,000 watts.

Radio's star of destiny shines bright. The clouds of the early days have been dissipated by the research laboratories. Radio is marching on. Television is ahead. Those who have faith in it foresee undreamed-of possibilities. They have faith because even the research engineers and scientists see no end to what may be accomplished. That is why a Radio City was founded. This electrical acropolis, in fact, the entire structure of broadcasting, has for its basis invisible waves, which according to the courts belong to no one because no one owns their medium. The broadcast license as issued by the Federal Radio Commission is granted for only six months. Yet there seems to be a feeling among the broadcasters that priority counts for something, and that is one reason why the pioneers are confident of the future; that is why they continue to expand and to invest further in the science and the art of ethereal entertainment.

While this radio center is to house four large theaters, one seating 7,000; a motion picture auditorium seating 5,000, another for musical comedy and one for legitimate drama productions, and possibly a great symphony hall, the builders are counting on, by means of the microphone and televisor, a greater audience numbering many millions. The entertainment in this radio city will find its way quickly to distant places, through broadcasts and television. The melodies will travel through space and will entertain also on the disk of the phonograph through electrical recordings in studios of this musical center.

NEW OPPORTUNITY FOR TALENT.—Culture, education and entertainment comprise the aim of the enterprise. It is

A CASTLE AND A CITY OF DREAMS 139

expected to do much to promote all the arts in the range of electrical entertainment. David Sarnoff, president of the Radio Corporation of America, foresees that artists will step upon the new variety stage and, with the developments promised eventually in television, entertain face to face a worldwide audience. He sees the dramatic and musical performances on the stage of these theaters flashing out to the countryside. He predicts that this Radio City will encourage creative talent, because of the vast facilities of expression. He expects a great advance in the service which entertainment and musical education can render the public, both in and out of the theater. In this city of music, technical and artistic development will go hand in hand toward new goals of progress in the art of communication and recreation.

The stage, the silver screen, the television screen, the phonograph, the microphone and all the avenues which radio entertainment travels will be brought together.

There will be twenty-seven broadcasting studios. All will be equipped for television. It was not so long ago that a radio studio twenty feet square was looked upon as large. It would easily accommodate a good-sized jazz band! But suitable dimensions of a studio are no longer judged by the number in an orchestra. Some of the new studios in the radio city will be two or three stories in height. They will be concert halls in effect, carefully planned for their acoustics. Each of the four big theatres will be designed for broadcasting. Actors will perform not only for the immediate audience but ultimately, perhaps, for the whole country. Ten of the twenty-seven studios will be equipped for photography and electrical recording. The public will be provided space so that they can see the radio entertainers at work. An Opera House is also planned.

The plans for the new studios are taking into account the fact that broadcasting, established upon a democratic basis in the United States, is not only a medium of mass entertainment, but that it has added to the cultural and educational values of modern life. With the great theatrical and musical enterprises to be created in this development, the broadcasting center of the country is being joined with the dramatic stage, with opera, with vaudeville, with talking pictures, with the symphony hall. Broadcasting facilities will be at the side of every artist whose performance can command a wide audience.

Nor will the talking pictures be neglected because of television. In fact, they, too, may travel on radio's wings at the same time they flash on the screen before the visible audience.

"Broadcasting at first seemed to be everybody's business," said M. H. Aylesworth, president of the National Broadcasting Company. "It was as though civilization had been waiting for a return to first principles, not only as to keeping in touch with leaders of the nation by spoken word, but also for entertainment. Here, at last, is a means of combining hundreds of thousands, even millions of listeners into a great forum. The proverbial four walls of the home, heretofore serving to isolate the family from the outside world, are now dissolved as the family takes its place daily in the forum of the air. Invited speakers—invited by a twist of a dial—musicians, educators and others come into the home from far and wide. Radio is the realization of a dream worthy of Jules Verne."

BRAND-NEW STAGECRAFT.—The traditional arts could not alone have brought about radio's growth, Aylesworth points out. It has been necessary to develop a special brand of showmanship or stagecraft, especially applicable to the microphone. In much the same way that the silent drama of the motion picture screen produced new problems in the histrionic art, so has broadcasting introduced new standards in musical art. The radio playwright has had to be developed with a special technique able to place the players in a mental setting, continually identifying them, and otherwise to make up for absent scenic effects of the presentation. The microphone's musical director has had to learn how to concentrate complete operas or musical comedies into the shortest possible time without impairing their worth.

WILL TICKETS BE SOLD?—It is doubtful if sound broadcasting alone could ever form the foundation for Radio City. Naturally, there are plenty of economic as well as technical problems to be solved before this huge entertainment center is functioning on a paying basis, unless, of course, some philanthropist takes it over. So it is no wonder that the listeners, who are apparently destined to become "lookers," are wondering what a key o ticket to this magic acropolis will cost or will the television performance be as free as the music in the air?

When broadcasting began in 1920 no one seemed to know exactly how far, or where, it was going. There had never been anything like it in history. But today broadcasting has enabled man to look further into the future. A great destiny is seen for radio and a new era of electrical entertainment. Those who are planning the television center foresee a radical change coming, in which every home in the land will be a theater in itself, linked by radio with this nucleus of entertainment from which music and television entertainment will flow into space. Radio vision will give the American public a powerful field glass through which those in Iowa, California, Texas, and other distant points can look through space, across the horizon and into the new temple of radio which will probably be completed in 1935.

PICTURES MIGHT BE SCRAMBLED.—Some are wondering how this big investment in Radio City will pay. How can a theater survive if the audience is not called upon to buy tickets? One theater in this capitol of radio will seat 7,000 and the talking-picture auditorium will seat 5,000. Tickets will be sold for these seats. But outside, on the other side of the televisor, is a countless audience numbering many millions. Will they get the same entertainment gratis? Of course, they must buy a television receiver. But will the television waves be scrambled so that no one can see them unless they buy a certain receiver designed to unscramble the waves which carry the entertainment? Not for a long time to come; it is difficult enough to scramble the voice and have it recognized without attempting to scramble smiles, tears and dramatic action.

Leaders in the radio industry, those who were building and selling sets as fast as the factories could turn them out in the early days to meet the urgent demand, objected to broadcasting operating on a toll principle. They opposed on the ground that "we must keep a free general system of broadcasting. The whole industry is founded on that idea in America. Broadcasts must be accessible to all."

If broadcasting had not captivated the public fancy so quickly it might have grown slower. A toll system might have been adopted. But by 1923 it was considered too late to introduce a secret system chiefly because millions of receiving sets, loudspeakers, batteries and vacuum tubes had been sold to the public. If a secret method of transmission had been applied, all the listeners would have had to scrap their receivers and buy new machines designed to operate as a key to unlock a mysterious combination of wave lengths. All of the transmitters would have had to be rebuilt. The radio industry would have been paralyzed and its growth retarded.

Hammond's "narrowcasting" invention, might have made feasible the collection of fees from listeners. This would have enabled the broadcasters to pay top-notch entertainers without being obligated to advertisers. It may have been too late to adopt the secret system in 1924, but today the time is opportune for the broadcasters to look ahead and adopt a

A CASTLE AND A CITY OF DREAMS 143

method, if they care to do it, whereby everybody cannot pick up a television show free.

The broadcasters contend that they are not worried, however. They know that radio performances as a free commodity attract the largest audience. If the program were broadcast on an almost unlimited combination of wave lengths, only those who pay for the "key" would be able to eavesdrop. What the broadcasters, who sell time, most desire is circulation. If they can convince a program sponsor that they reach an audience of 20,000,000, the advertiser is more likely to buy time than if the audience is restricted to 500,000, limited by a secret system. The broadcasters are looking ahead to television as a great boon to national advertising. Whether they would adopt a toll idea is extremely doubtful. They are not anxious to limit the size of the audience by means of a mechanical contraption. The outlook is that advertising will support television just as it does broadcasting.

THE THEATERS WONDER.—It is possible, but not altogether probable, that some day an inventor will discover how to stretch a "high wall" around some parts of the television show. Already theatrical producers are wondering how they could afford to let a show be televised.

How many would travel to Palmer Stadium to watch Princeton play Yale if they could sit comfortably at home and see the football game at a television screen? Would 75,000 gather from all sections of the country to see the World's Series if a television eye gave the nation a grandstand seat free? Would Madison Square Garden be packed to capacity for a championship bout if a television eye hovered above the ringside to send the scene across the countryside? And the television eye would be so located that no seat in the house would afford a finer view. The lookers-in on the radio would probably see more than the majority in the arena. Television receivers might be rented in much the same way that the telephone system is handled, but that is doubtful because radio broadcasting has established a precedent not easily changed.

ONWARD TO THE PARLOR.—David Sarnoff is sure that progress in the electrical arts inevitably points to millions of little theaters added to the constellation of entertainment already made possible by radio, talking pictures and the modern phonograph.

"A separate theater for every home—although the stage may be only a cabinet and the curtain a screen—is, I believe, the distinct promise of a new era of electrical entertainment," said Sarnoff. "The stage, the concert hall and the opera first entered the average home with the phonograph. It is true that musical instruments in some form have existed since the dawn of civilization, but with the exception of the first crude piano rolls, it required the creative artist or the amateur to make them vibrant with music. The phonograph reproduced music and speech wherever it entered. It gave to the home the recorded art of the concert performer, the operatic star, the stage favorite.

"Now comes the promise of television as applied to the theater of the home. Important as has been our progress in the development of sight transmission, great technical problems still remain to be solved before such a service can be established upon a practical basis.

"Television will be harnessed to the motion picture screen so that a great event might be simultaneously recorded in a number of key cities throughout the nation and the talking motion picture film distributed again by television to millions of homes some hours after the actual occurrence. Television, when it does come upon a practical service basis, promises to supply a vast invisible channel of distribution for motion pictures in the home."

CHAPTER TEN

VAULTING ACROSS TEN YEARS

It is November 2, 1930. Radio broadcasting is celebrating its tenth anniversary. It was ten years ago today that the first program went on the air at Pittsburgh. The first decade of broadcasting has established an industry. It has entertained, informed, educated and employed thousands of people. The second decade is beginning.

What will happen in the next ten years? What new wonders will the mushroom-shaped cathode ray tube and glowing neon lamp with its noble gas achieve? Will the scanning disk survive?

SURPRISES ARE PROMISED—NOVEMBER 2, 1930

Several days ago Dr. Alfred N. Goldsmith sat at luncheon in the Hotel Astor, and as he looked out on Broadway his eyes appeared to miss the crowds, and the hustle of noonday traffic. He seemed to be looking farther into the distance. His mind was focused on the future of radio. That was the subject under discussion.

"Think of it," he said, "ten years have shot by since broadcasting started. Ten years ago radio was a mere infant. How it has grown! Today radio is a world-wide and mature institution. We are on the threshold of another wonderful decade. It is uncanny to imagine what radio will be like in 1940. We are entering a new era of electrical entertainment."

Why? Because the radio pioneers blazed a splendid trail in broadcasting. In a brief span of years they have established engineering and artistic precedents of basic importance which have enabled the building up of mass communication by radio telephony into a great industry. During the last few years the technique of broadcasting has been refined and the scope widened until, today, in 1930, it stands as a highly developed and universally accepted form of major entertainment supplied to the people of the world.

"It is but natural to ask whether the amazing rate of progress during the last ten years can be maintained, and whether 1940 will see radio as far improved compared to the present-day conditions as is the broadcasting of today when compared to that of 1921," said Goldsmith. "To the public, which is already well satisfied in the main with the excellent performance of the better modern receivers and transmitting stations, it would offhand appear as if progress from now on would be slower than in the past. Yet this theory is extremely doubtful, and the scientists and engineers have every reason to believe that not only electrical entertainment in general, but also radio broadcasting in particular, will improve in performance, convenience and scope, and at a marked pace, as the years go on. New principles and methods, as yet only in the minds of the inventors, or at best in the laboratory, appear to beckon the radio art forward to new accomplishments and triumphs.

IT IS 1940!—"And so, vaulting over ten years, imagine we are in 1940. Looking about at the field of electrical entertainment, what do we find?

"We enter the radio broadcasting studio of 1940. The microphones are nowhere in evidence for the methods used so successfully in 1930 for sound motion picture production, with remote and concealed microphone, will have found their place in broadcasting. Devices oddly like cameras will point at the actors, picking up their images for television transmission, perhaps in color. Motion picture cameras are in evidence. The studio, with its special backgrounds and furnishings, will look much more like the stage of a theater or a motion picture studio than like the orderly room which it resembled in 1930. Television pick-up men and camera men, sound recordists and control room experts are busily at work. Actors troop out of their dressing rooms in the costume suited to their performance. Their words and their appearance are carried instantaneously by wire line or radio connection to a multitude of outlet stations.

"In the control room, provision is made in the case of the more important broadcasts to record both the picture and the sound of the performance, either on photographic film or on some equivalent material. The cameras are taking pictures of the television performance which is being broadcast. Thus, the public can purchase sound motion picture records of any particularly attractive or historically important broadcast which has been presented. School children and their parents will have the advantage of seeing and hearing historical events which have been recorded for them at the same time as they were broadcast.

MAN's NEW SERVANT.—"Entering the living room of 1940 one might judge from the preceding description that all the electrical entertaining devices to which reference has been made would prevent the owner of the home from entering the living room because of the congestion of the pieces of furniture. Yet such is not the case. Instead of several cabinets each containing a single instrument, the electrical entertaining equipment is assembled in relatively few cabinets and in some cases even in a single cabinet known as the electrical entertainer. Essentially the electrical entertainer requires only two outlet portions, namely, a screen for showing a picture and a loudspeaker for producing a sound. Back of the screen is arranged either the television projector or the sound motion picture projector, or both. The educational and entertainment possibilities of such a device are limitless.

"In 1940 we have the electrical entertainer at the disposal of the public. Its significance in the stimulation of musical taste, as an incentive to the creation of music at home, as an entertainment device and as a means of education has, it is believed, opened a new era. The electrical entertainer has already become a part of the life of the world," Goldsmith declared. "If we now look forward to 1950, some of its capabilities will have been further explored and mankind will have begun to derive a larger measure of the inestimable benefits which the applications of electricity can bring to it. And so, through the decades, the force which first frightened man when it flashed in the lightning and roared in the thunderbolt will not only become his servant but even his ally in improving his mind, broadening his cultural taste, and brightening his hours of leisure."

CRITICS LOOK AT TELEVISION.—Luigi Pirandello's *The Man with a Flower in his Mouth* is televised in Baird's London studio while dramatic critics apply their eyes and ears to the sights and sounds that come to them by radio. Station 2LO handles the sound part of the performance on the 356-meter wave, while a regional station broadcasts the images on 261 meters.

The critic of the London *Times* remarks that the difficulties already overcome are many and remarkable, but "let it be admitted at once that plays by television are as yet a subject for men of science and not for critics of the finer points of acting."

It is estimated that approximately 1,000 television receivers are being operated in England. Baird, the inventor, hibernating in his isolated laboratory atop Box Hill, twenty miles from London, is reported to be well along with a new television system which is radically and fundamentally different from the usual practice.

TELEVISOR HAS FOUR PARTS.—The televisor now being used in London is described as having four essential parts:

VAULTING ACROSS TEN YEARS

the graduated scanning disk, driving motor, synchronizing mechanism, and the neon lamp. The scanning disk is twenty inches in diameter and has thirty accurately cut circular holes arranged in the form of a spiral. The first and last three holes in the spiral are cut square. This results in greater detail at the center of the screen than at the edges, and is called "graduated exploration."

The automatic synchronizing device has two small control knobs on the front panel of the television receiver. One of these knobs must be adjusted until the image as viewed through the lens is brought to rest. Manipulation of the other knob adjusts the image to the correct height, so the bottom of one face and the top of another are not seen at once.

The first operation is to tune the receiver. A loudspeaker can be used as an aid. When the reproducer emits a shrill note that means the image is being intercepted. The televisor is then switched on. The observer sees streaks of light that slant from one side to the other. The synchronizer knob is turned until the streaks appear horizontal. When exactly horizontal and in tune with the sending station the images appear on the glass screen. The picture may then be framed to please the eye, that is, it can be centered and clarified by the proper tuning.

HEAD AND SHOULDER VIEWS.—Television spectators are at present restricted to the reception of head-and-shoulder and other small images. Large outdoor scenes cannot be broadcast at present, and the engineers explain that whether the large scenes are likely to become practical within the next few years is a question in which the whole future of television is bound up. It is true that considerable depth has already been achieved in the transmitted image, but it lacks the full stereoscopic effect for which the engineers are striving. They are proud that television in color has been demonstrated experimentally. Bouquets of red carnations, blue delphiniums, also strawberries amid their green leaves in white baskets have been seen on the radio in striking effects. But color reception calls for a special receiving set and tubes. Neon tubes have been used in European experiments for "painting" the reds, while mercury vapor tubes handle the blues and greens.

BLONDES ARE PREFERRED.—Every person and every thing does not necessarily televise well. The clarity of the image depends to a great extent on whether the face which is being sent is a good "television" countenance. The London televisor at its present age apparently prefers blondes.

Why brunettes appear to be in disfavor is a riddle. It seems that the ideal television face is round and smooth with shallow lines and the fewer hollows the better. Faces differ for television work as they do in film work, and it may not be long before a recognizable type of television face emerges from these pioneer broadcasts. A minimum of make-up is used because the televisor will produce a pasty image if too much make-up is put on.

Facial gestures are encouraged. They give life to the image. No spotlights are used in the studio. The vocalist at the microphone is in almost complete darkness with a small red light at one side to indicate the position of the microphone, while a flickering white light beam scans the face. That is one of the recent improvements. Heretofore the person being televised was obliged to sit under a light of blinding brilliance and so hot that it peeled the skin from the forehead in the course of a long sitting. But the latest machines, radio cameras and lights involve not the slightest discomfort to the person under the stare of electric eyes.

PIN POINTS OF LIGHT IN CALIFORNIA—DECEMBER 14, 1930

Philo T. Farnsworth, described as a modest young man who can apply basic theories in a common sense way, visits New York having first made a call on the Federal Radio

VAULTING ACROSS TEN YEARS

Commission in Washington. He announces that he has succeeded at his California laboratory in narrowing the wave band required for clear television pictures. Advantages of electrical scanning rather than a mathematical formula pertaining to wave lengths have helped him in this work.

Farnsworth reports that he has developed a cathode ray tube which together with pin points of light, will eventually make television commercially practical. He says that this television instrument can be used in conjunction with existing broadcast receivers. The tube is about the size of a quart jar and the picture appears on the bottom of it.

"I have abandoned the old idea of a whirling disk with its motor and other contraptions," said Farnsworth, in describing his system, which first found its way into the newspapers in October, 1928. A simple beam of light does the trick. The entire receiver including a cathode ray tube and its power unit, can be housed in a box slightly larger than a foot in dimension. It is plugged into the broadcast receiver following the detector tube. If the cathode bulb burns out the owner releases a catch, unscrews the tube like changing a light bulb, and inserts the new one. The entire device and tube equipped for use with a broadcast set should cost less than a hundred dollars. The flat end of the cathode tube takes the place of the grille work which ordinarily covers the loudspeaker opening.

"In the laboratory at the present time I have a system in operation which requires a wave band only six kilocycles wide to carry the images from the transmitter to the receiver. It is possible to reduce this wave band to five kilocycles so the pictures can be sent out by regular broadcasting stations. I believe that television will be combined eventually with sound programs over one ten-kilocycle channel by placing the music or voice on one side of the carrier wave and the image on the other side."

TUBE CALLED "DISSECTOR".—Cathode ray tubes are used

for both transmission and reception in Farnsworth's system. The tube at the broadcasting station is called "an image dissector bulb." It is a high vacuum, cold cathode type of tube described broadly as a photoelectric cell designed so that "an electron image" of an optical image is focused on the cathode surface through the flat window opposite it.

If a fluorescent screen is placed in the path of a photoelectric cell or target electrode, the original optical image is reproduced. For this to happen, however, it is essential that every electron emitted from any point on the cathode surface must impinge on a corresponding point in the plane of the electron image. The cathode rays have a tendency to spread. Therefore sharp focusing of the electron image is important for successful pictures. By applying a magnetic field of the proper intensity the image is focused in such a way that the lines of force are parallel to the axis of the bulb. The image can be shifted by two transverse magnetic fields, so that the entire picture can be moved across the aperture in the target shield, thereby achieving a zigzag scanning of the image. This is known as electrical scanning. No whirling disk is required.

AN ELECTRON GUN SHOOTS.—Farnsworth calls the cathode ray tube used at the receiver, an "oscillite." It transforms the incoming picture impulses into a visible image. The scanning at the receiving end is carried out by means of two sets of coils mounted at right angles to each other, just as at the sending end. There is an electron gun element designed to drive the greatest possible number of electrons through an opening so that the beam can be easily focused. Synchronism between transmitter and receiver is achieved by the use of two alternating currents of saw-tooth wave form generated at the receiver, identical with those at the transmitter. These currents are made to induce a strong voltage into the picture frequency circuit during the steep part of their slope. These pulses are utilized at the receiver to hold the local generators in step. And the pulses, which are transmitted only during the interval between individual pictures, also serve to turn off the oscillite spot during the return part of its path. The main advantage of this system is that no extra communication channel either wire or radio is needed to convey the synchronizing impulses, nor is additional apparatus required.

The inventor calls attention to the fact that the sawtooth wave form of alternating current is employed for energizing the coils because if a sine wave current (one that rises and falls rhythmically), were used a double picture would appear at the receiver, whenever the two currents were not in phase. Each scanning frequency at the receiver is generated by means of a helium glow discharge tube in combination with a small power tube employed as an oscillator and one stage of amplification.

THE NEON TUBE IN ACTION.—Now let us turn to the mechanical method of scanning to see how the neon lamp performs its duty. It is to the television set what a loudspeaker is to a sound receiver.

Suppose you are looking into the aperture of a television set equipped with a disk similar to one at a transmitting station and also provided with fifty small holes arranged in spiral form. A motor revolves the disk at the same speed in exact synchronism with the disk at the sending station. The observer looks at a small rectangular opening or frame in front of the disk. This frame is of such dimensions that only one hole on the rim of the disk can appear in the field of view at a time. As the disk whirls, the holes pass across the frame one after another in a series of parallel lines, each displaced a little from the preceding one until in one revolution of the disk the entire field has been covered.

GAS RULES THE COLOR.—Beyond the disk is the neon glow lamp. It contains two elements sealed within a glass bulb that contains one of the so-called noble gases, such as neon, argon or helium. The cathode is a flat metal plate of shape and area sufficient to fill entirely the field defined by the frame in front of the disk. The positive electrode or anode of this lamp is a similar plate separated from the cathode by about one millimeter. At the proper gas pressure this tiny space between the plates is within the "cathode dark space" where no discharge can pass. As a consequence, the glow discharge develops on the outer surface of the cathode, where it shows as a perfectly uniform, thin, brightly glowing layer. The color of the light depends upon the gas used. Neon produces an orange or pinkish hue. Argon properly mixed with nitrogen gives a white light. Helium gives a blue-white light but requires higher voltage to produce satisfactory ionizing. Neon was first used because the eye is more sensitive to orange than to white so that the images appear brighter in that tint. Furthermore, the voltage necessary to ionize neon is comparatively low.

Now, as a hole in the disk moves across the field, the observer looking through at the neon lamp behind the disk sees the aperture as a bright spot. Each spot is on the plate of the neon lamp for a mere fraction of a second. When the disk is rotated at high speed, the observer, owing to persistence of vision, sees a uniformly illuminated area in the frame, provided a constant current is flowing through the lamp. The brightness of the neon lamp is directly proportional to the current flowing through it. When a picture is being received, the lamp is operated directly from the incoming picture current. As a result, there is at any instant, in the field of view at the receiving station, a small aperture illuminated proportionally to the brightness of the corresponding spot of light on the distant subject being televised. Therefore, the observer sees an image of the distant person reproduced in the frame at the receiving station. The image on the plate of the lamp is usually about an inch square. Lenses magnify it for the screen.

This flat-plate neon tube has disadvantages, chiefly, that it diffuses the light whereas it would be far more efficient to concentrate the glow into a beam thereby obtaining a vast increase in the illumination of the picture. Then it would have more detail and larger size. But with the flat-plate tube only a fraction of the light gets through the scanner to the lens.

WATER-COOLING HELPED.—The search for a more intense light led to the development of the air-cooled and water-cooled crater neon lamps, which produce strong concentrated light instead of a diffused glow. The intense glow appears in a tiny hole the inside of which is coated with a mixture of calcium, barium, strontium oxides that emit electrons at comparatively low temperatures.

It is essential that the glow discharge lamps for television contain neon, argon or helium, because these gases produce a light that can be modulated with sufficient rapidity to trace the incoming radio signal. As television developed it was found that water-cooling enabled the use of higher currents, which resulted in greater illumination. It was also discovered that if a small amount of hydrogen is mixed with the neon the active life of the lamp is extended. Some neon bulbs are so designed that hydrogen can be fed in through a valve. This is done periodically when the action of the lamp becomes sluggish and the image fuzzy.

There is another type of glow tube known as the crater-mercury vapor lamp. It emits a blue-white light through a pinhole in the center of a metal disk inside the tube. A drop of mercury vaporizes when the current is turned on, and causes a white instead of the pink light characteristic of the neon bulb.

"The uniformity of the glow of neon tubes and the sputtering from the active surface depends on the use of the proper technique in preparing the cathode surface," said W. H. Weinhart of the Bell Telephone Laboratories. "Sputtering is the dislodging of material from the surface by impact of ions from the glowing gas. The matter released leaves the cathode's surface with high velocity and deposits on the inside of the bulb directly in front of the glow. This soon renders the lamp useless by reducing the intensity of the light as viewed through the bulb. It has been found that beryllium sputters far less than other materials and, therefore, is used for the final plating of the cathode. Beryllium is not easily worked. It can neither be electroplated nor readily deposited by cathode sputtering; therefore, it is necessary to deposit it by vaporization and condensation. This is done in a high vacuum to prevent oxidation and to leave the surface as free from gas as possible."

And so these mute glass bulbs blink and glow as electrons work miracles within the thin glass walls, painting pictures with invisible crayon-like points, annihilating space so that the human eye can see distinctly, much farther than just across the street.

PART IV

THE CALENDAR TURNS AGAIN

Chapter Eleven

TELEVISION TECHNIQUE AND ARTISTRY

In the vanishing days of 1930 there are broadcasts that reveal as never before radio's international influence. America on Christmas morning eavesdrops on melodies from Japan. It listens to the 400-year-old bell tolling in the tower of the Cathedral of the Immaculate Conception in the Philippines. It tunes in Hawaiian guitars being strummed in Honolulu. And later in the day a church service from London crosses the sea, and Germany's musical greeting wafts across the American continent.

Then again on New Year's Day, Italy joins the worldwide circle of friendship. Premier Benito Mussolini at his desk in the Palazzo Venezia speaks into an Italian microphone, a pledge of peace and goodwill that echoes round the globe on the wings of radio projected into the air of the Eternal City.

Surely if sound can thus girdle the earth, television cannot be so far away.

It is 1931!

Electric Eyes That "See" Red—January 11, 1931

Electrical research has pushed television nearer to the talking motion picture in clarity and simplicity. The cumbersome, heat-creating electric arc light heretofore used in the majority of television transmitters has been replaced by a powerful incandescent lamp. And the neon bulb, instead of casting a pale orange glow, now throws a more powerful beam of light to paint clearly the image or scene on the screen. No longer are the neon rays scattered and feeble.

The rim of the whirling disk has been fitted with seventy-two sensitive lenses that concentrate the neon tube's light, thereby giving the picture greater contrast. These progressive steps have enabled the engineers at the Bell Telephone Laboratories to build a television set of half the size. They are proud that television images really have what they call "definition."

They have developed a new cæsium photo-cell that "sees" red. It detects the red pigment of the skin and makes the image more life-like. It does not dispense with "eyes" of the potassium variety, which are sensitive to blue at the other end of the spectrum. But when potassium cells are used alone the face is more likely to be blotched and darker than normal. Now the images appear on the screen more nearly as if seen face to face in daylight.

A. R. Olpin, engineer of the electro-optical research division of the Bell Laboratories, is credited with much of the development made possible by the cæsium cell. He perfected it for television, so that the lips would look natural and the eyes clear. The ears no longer look "dead" white. They have a tone or shading as the lights play on them.

The new high power incandescent lamp avoids the flickering always present to some extent in an arc. Therefore, the image is steadier. The maintenance and adjustment of the incandescent lamp is simpler. A further advantage is that the incandescent bulb's filament, operating at a lower temperature than the arc, radiates more light at the longer wave lengths (red light). This facilitates improvement in the scanning system.

EYES INSENSITIVE TO BLUE.—At the first two-way television demonstration in 1930 the scanning beam was filtered to pass only blue light, and the photoelectric cells (potassium-sulphur-vapor) were sensitive chiefly to light in the blue region of the spectrum. With two-way television it is necessary for each person to see and be seen at the same time. Therefore, each speaker must be scanned by a beam of light while looking at the images formed by the neon lamps. The light from the neon tube is not of high intensity. Therefore, its effectiveness would be decreased if the person being televised were flooded by a strong light from some other source. He would be blinded and could not see the image from the other end of the line. The human eyes, however, are insensitive to blue light. By giving the scanning light a bluish tinge it has small effect on the ability of a person to see the received neon image as it appears on a screen in front of him in the two-way television booth.

The effect of using only blue light, however, was to make the yellows and reds in the face too dark in comparison with the whites, such as a linen collar. This is because little blue light is reflected from yellow or red surfaces. To secure greater naturalness in the image a deep red component has been incorporated in the scanning light beam, making it purple instead of blue. Two photoelectric cells of the cæsium-oxygen type have been included, which are extremely sensitive to red light. The result of this scanning from both ends of the visible light spectrum is to produce an image that is a more faithful reproduction of the original. The effect is much like that which would be obtained by scanning with light from the middle of the visible spectrum; the definition of certain important points, such as the eyes, is distinctly improved.

The cæsium cells are only about half the size of the potassium cells, but because of their high sensitivity to light of long wave length (red rays) and to the richness of the incandescent lamp in light at the red end of the spectrum, two cæsium cells are about as effective as the twelve potassium cells that supplement them.

EYES NOW SEE ALL COLORS.—In outlining the progress being made in photoelectric cell manufacture, Olpin said that electric eyes can now be obtained which are sensitive to almost any color.

The new neon tube is known as the "crater type." It has a much smaller metal plate inside the glass envelope than the former bulbs of the plate type. It glows with an intense orange light generated on the surface of a small electrode, which is slightly concave like a reflector. The effect is to cause the neon beam to be projected in a thin pencil of light through a series of lenses which further concentrate and direct the beam to a series of seventy-two tiny lenses placed in spiral form around the rim of the receiver's scanning disk.

The lenses are arranged so the full power of the beam is deflected over the entire area of the screen on which the observer sees the intercepted image. The result is far greater contrast between the light and dark areas which make up the picture. At the beginning of each revolution of the scanning disk the orange light beam is painting a dot in one corner of the screen. Before the light sensation has died away in the eye of the observer many thousands of such dots of illumination have been painted, flooding the entire screen with light. All this takes place in one-eighteenth of a second. Eighteen of these pictures per second, resulting from the Nipkow disk turning a like number of times in one second, deceive the eyes into seeing a smoothly changing picture.

INCANDESCENT LAMP AIDS.—Some idea of the efficiency of this new television optical system is presented by a comparison of the arc lamp formerly employed to illuminate the object being televised and the new incandescent lamp, which is but slightly larger than a 100-watt light bulb.

The arc light emits illumination of about 18,000 candle power, whereas the new bulb produces but 2,000 candle power, which is more easily controlled and concentrated. About 900 watts of electric energy pass through a filament less than two inches long within the tube. It is one of the most intense lights in existence, rivaling even the light of the larger and more complicated arc lamp.

When the arc light was used the television machine had to be shut down frequently so new carbon rods could be placed in the holders. Now this is unnecessary. When an incandescent lamp burns out it is unscrewed from the base and a new one inserted in a few seconds. Formerly the machines were stopped during all minor adjustments. Now the incandescent lamp keeps the television machine in running order for more than 200 hours of operation.

"Ghosts" That Hover in Space—February 22, 1931

Television images and their phantom "ghosts" are playing tag around the skyscrapers on Manhattan Island. Research experts equipped with sensitive receivers are following them to watch their antics. The engineers are trying to discover what wave lengths are best for dodging the ill effects caused by buildings that touch the clouds. And they are beginning to wonder whether or not it will be more prudent to erect the television stations outside the city limits where many of the broadcasters are located.

Bounced Back from the Sky.—Some of the images reach the screen as an apparition. They no sooner flash into view than the same face reappears in a faint shadowy form, a specter returned from the infinite. There may be three or four of these sprites. The experts know what causes them and they would like to find a way to erase them from space.

It seems that the "ghost" might be bounced back from the Kennelly-Heaviside surface, the earth's blanket of electricity-conducting air about 100 miles up in the sky. Its altitude is ever-changing, billowing up and down like the top of a circus tent in a gale. Radio waves strike this layer, sometimes called a radio "mirror," and are reflected back to earth again. Shifts in the height of the layer send the waves back at varying angles and as a result the waves are not always perfectly synchronized with the other waves that travel along the ground. That causes fading, blurred pictures and double images. Or the waves might travel out to a mountainous region to be reflected as a sort of echo. Scientists have observed that these radio echoes have come back from distances beyond the orbit of the moon. The mileage is determined by noting the time lag between the main signal and the echo. On the other hand, part of the transmitter's energy may follow a sky-wave route while another portion travels along a ground wave. Both waves do not always arrive instantaneously at the receiver because their paths differ in length. That is another cause of ethereal "ghosts."

"One night at nine o'clock I noticed a 'ghost' flashing by," said Sanabria. "It grew more violent until eleven o'clock after which it was seen, but fainter. Finally it vanished about an hour before sunrise. I was watching it about nine miles from the transmitter. Incidentally, an antenna in the open picks up much less 'ghost' than one shielded by a steel building. This suggests that the ground wave is considerably weakened by absorption. The sky wave is much stronger. I have attempted to observe the correlation of visible clouds with 'ghosts' that sometimes appear in the afternoon, but often the two occurred simultaneously, and at other times when 'ghosts' appeared only a few clouds were in the air, so they could not be blamed for creating the sprites."

METAL "UMBRELLA" IS TESTED.—Experiments are being conducted by the Bureau of Standards to exterminate these "ghosts" that stalk the airways. One idea consists of placing a large metal sheet or "umbrella" over the transmitter's aerial. This sheet absorbs all the sky waves or refracts them toward the ground before they can emanate far from the sending station, and therefore, the ground wave is sent out alone. Although it is possible to eliminate the double image in this way, signals broadcast under such conditions cover only a short distance. It is the sky wave that travels farther and remains strong longer.

When the metal sheet is used above the aerial, and signals are sent out on frequencies between 43,000 and 80,000 kilocycles, the weaker waves are absorbed and a beam-like wave emanates intensely. Although the tests have not been carried far enough to show definite results, it is possible that the idea may be developed as a solution of the problem.

Television Moves Nearer the Home—March 11, 1931

Seeing by radio is being brought definitely nearer to commercial development by research and technical progress made in the laboratories, according to a report issued by the Radio Corporation of America, which has a corps of experts led by Zworykin and Alexanderson working night and day to perfect television devices for the home. The report reads:

> Public interest in the new service promised through sight transmission by radio, and the new industry which the manufacture of television sets for the home now brings into view, requires a precise statement with regard to its developments. It must be recognized at the outset that while intelligence may be transmitted through either the ear or the eye, the services which radio may render through sound and vision do not compete with one another. Each has its peculiar and distinct function.
>
> Sound broadcasting, upon a continually rising scale of public interest, is engaged in developing its major possibilities. Similarly, the sound equipment industry continues to be subject to further development technically and industrially. Sound broadcasting and sound reproducing equipment constitute a distinct division of the radio art.
>
> While television during the past two years has been repeatedly demonstrated by wire and by wireless on a laboratory basis, it has remained our conviction that further research and development must precede the manu-

facture and sale of television sets on a commercial basis. In order that the American public might not be misled by purely experimental equipment and that a service comparable to sound broadcasting should be available in support of the new art, we have devoted efforts to intensive research into these problems, to the preparation of plant facilities and to the planning of studio arrangements whereby sight transmission could be installed as a separate service of nation-wide broadcasting.

It is felt that in the practical sense of the term, television must develop to the stage where broadcasting stations will be able to broadcast regularly visual objects in the studio, or scenes occurring at other places through remote control; where reception devices shall be developed that will make these objects and scenes clearly discernible in millions of homes; where such devices can be built upon a principle that will eliminate rotary scanning disks, delicate hand controls and other movable parts; and where research will make possible the utilization of wave lengths for sight transmission that will not interfere with the use of the already overcrowded channels in space. . . . Progress already made gives evidence of the ultimate practicability of a service of television.

Midst Lightning and the Rain—April 26, 1931

The actors in a première television performance go forth from the aerial wires of station W2XCR, New York, to battle with lightning flashes. The images of Felix the Clown, Gertrude Lawrence, Dorothy Appleby and a host of others are subjected to a severe test on their first ethereal flight. Atmospheric conditions are bad. It is no day for timid images traveling on the wings of feeble radio waves to stray far from home if they are to maintain their identity. Nature has sent a downpour of April showers accompanied by lightning flashes and the roar of thunder. And lightning, the mother of static, is never kind to radio images. It freckles them, cuts away part of the countenance, streaks the face

TELEVISION TECHNIQUE AND ARTISTRY 167

and makes the image look like one of the ghosts and witches that Macbeth saw midst the lightning and the rain.

FIGHTING AN ELECTRIC BARRAGE.—But despite the electric barrage, severe enough to silence the powerful voice of WEAF for more than an hour, the images that jump off the aerial wires atop 655 Fifth Avenue are plucked from space in Baltimore and at observation outposts in the metropolitan area. It is possible that they traversed greater mileage, but television sets are scarce today compared with the millions of broadcast receivers.

It is estimated that there may be about 200 television receiving outfits in the New York area. Many of them are home-made, and are owned by amateur experimenters, as were the early broadcast receivers in 1920. Chicago is believed to have from 500 to 1,000 vision sets, chiefly because there has been greater activity in visual broadcasts in that area.

The images broadcast by stations W9XAO and W9XAP, Chicago, have been seen as they sped across the corn fields of Iowa, across the wheat fields of Minnesota and Kansas. They have found aerial wires in Michigan, Ohio and Missouri. One observer in Arizona reports that he caught a fleeting glimpse of them after they had traveled through the desert air. These television spectators say that they enjoy boxing bouts and performances that feature plenty of action.

HOW FACES ARE DISSECTED.—The miracle of television is realized when one stops to consider the process and the electrical surgery to which the persons televised are subjected. They must be scanned, that is, dissected. They are converted from light into electricity by wondrous eyes so sensitive to light that they change the lights and shadows into electricity corresponding in intensity to the original light pattern. Then the electrical impulses, to which the images are entrusted, are fed into a short-wave broadcasting machine.

It squeezes the face into electricity so that it can run up

the lead-in wire and out onto the aerial runway. It is no longer a face to be seen or recognized. It is a radio wave. The ether, or whatever that mysterious substance may be that occupies all space, is set in vibration. Then the images of Maria Gambarelli and Patricia Bowman dance across the skyscrapers and off across the countryside in the form of an invisible wave. A scene from the *Silent Witness* brings Lionel Atwill, Sylvia Field and others before the television optics so that they, too, may dart through the tall buildings to find slender antenna targets that bid them welcome to reappear, reincarnated on lenses and screens.

The faces squeezed into a flow of electricity at the aerial now run down the lead-in wires, showing no partiality in the selection of a home. All that is required is a television receiver to convert the invisible wave back into electricity and then into light so that it can be seen. And the wonder of it all is that, after this complicated operation, the faces again smile, wink and talk just as they did in the studio a fraction of a second before.

Those who look into the sky see no evidence that the forms of people—singing, dancing, acting and joking—are flashing through space, penetrating buildings, walls and even the human body, at the speed of light. Such is the wonder of the age that empowers a station to send out sound on a 254-meter wave, while its associated visual transmitter handles the images simultaneously on the 147-meter channel.

THE ART OF MAKE-UP.—Despite the fact that the Federal Radio Commission still contends that visual broadcasting must be pursued on an experimental and not a commercial basis, the broadcasting organizations are becoming intensely interested in learning more about the technical aspects and the technique of showmanship. Station W2XAB has been licensed to go on the air in New York and preliminary tests are being conducted to reveal what make-up is needed. A variety of other essentials are to be studied so that, when the Federal authorities pronounce television ready for commercialization, the broadcasters will be prepared.

"Our tests prove that a platinum blonde registers best on the television screen," said A. B. Chamberlain, chief engineer of the station in a report to the Institute of Radio Engineers. "So far as color being used in television make-up any results obtained could be duplicated with varying shades of black and white, but on account of black lipstick being objectionable to performers, a brown shade has been substituted with satisfactory results. We have also discovered that a rough white powder reflects more light from the face than fine powder, creams or grease."

It has been observed that the contour of the face has a great deal to do with the variety of the picture on the screen. The light striking a flat-faced person is better reflected to the photoelectric cells than from a person having deep-set eyes, pronounced cheek bones and sharp declivities. The reason for this is that the light striking any surface in a plane parallel to the beam of light cast on it from the scanner is thrown toward the floor or ceiling and not directly to the photoelectric cells. It is this fact that causes whiskers to appear at times when actually the person being televised has no beard or mustache.

"In television-cartooning it is best to keep the easel close to the scanner using a short focus lens to cover the desired area," said Chamberlain. "The light from the scanner to the object loses little of its intensity. On the other hand, the reflected light from the object to the photo-cells is attenuated as of the square of the distance. In this way a large surface close to the photo-cells gives a greater intensity than a small area at a greater distance. The same rule holds good for football boards, paintings, art objects and various types of lessons where books are shown. It is sometimes advisable to employ a longer focus lens to cover a small fraction of the picture in order to show more detail.

"We have found that shiny surfaces register poorly, because if curved they form a minus focus and if flat they add to the focus of the scanning lens throwing the light in only one direction, thereby striking only one cell. This results in an unbalanced picture generally containing straight streaks of black and white with very little semblance of the object itself."

ENGINEERS ACT QUICKLY.—Television requires a different type of control operator than does sound broadcasting. The operating engineer must have a good knowledge of arc lamps and their operation. This in itself calls for considerable skill; in fact, it is more than three-quarters of a moving picture operator's work. The scanning engineer must be able quickly to select the proper focus lens for the subject he is scanning and he must bring it into correct focus without delay. If there is any delay the observer has much the same reaction that an off-note would have on the broadcast listener. Therefore, to facilitate handling the lens a special six-lens turret is mounted on the scanner where each lens is kept at almost its correct focus. This turret will immediately swing the desired lens into position and little time is lost in getting a sharply defined picture. The scanning mechanism at W2XAB is also mounted on a pedestal to expedite a quick move in any direction, vertical or horizontal. It is important, however, that the heavy mechanism of the scanner be in exact balance in order to follow the movements of dancers, prize fighters and other action scenes.

"We have found it necessary to experiment with various backgrounds and to use them at varying distances behind the artists, and at varying distances from the photo-cells," continued Chamberlain. "Different drop curtains on rollers are mounted on a track system which can be quickly and easily adjusted back and forth from the scanner.

TELEVISION TECHNIQUE AND ARTISTRY 171

"It has been difficult to explain to people having a knowledge of photography, that it is not so much the color of artists or the background but their relation to each other so far as contrast is concerned, in television. They seem to cling to the idea that the photo-cells pick up an impression much the same as would a camera's sensitized plate. This is not the case, because the current flowing through the photo-cell circuit has a definite value with a given amount of light. This amount of current changes as the light varies. For instance, if we were scanning a pure black non-reflecting background, the image seen in the monitor would appear to be the same as if we were scanning a pure white background because there is no variation in the light that influences the photo-cells.

"While the cells are more sensitive to red and blue, nothing is gained by employing a red or blue background, because either of these colors would absorb all the white light thrown by the scanner, thereby lowering the efficiency so far as useful illumination is concerned."

DIFFICULTIES ARE MET.—Light cannot flow freely in a television studio, therefore, the artists often find it difficult to see and to follow their music or script. They constantly call for more light. But the engineers must refuse, because the more light in the studio the more difficult it is to get a definite variation due to the flying spot having to scan a subject already partially illuminated. Furthermore, as the studio lighting is increased the photo-cells have to work at a higher point of their characteristic curve, until finally the studio lighting is sufficient to start ionization which completely ruins the picture as well as the cells. To overcome this, the engineers keep the light at a minimum. They also use a yellow filter, because the cells are less sensitive to yellow than to any other color. They are working on a so-called parallel ray lighting system which will have little reflection from the surface of the music or script to be illuminated.

Another difficulty is the variation in focus between long and close shots, according to Chamberlain. For example, the focus may be at a maximum to pick up a subject ten feet from the scanner, when some one with a white dress shirt suddenly steps in as a close-up. That immediately throws an exceptionally heavy signal through the circuit. It trips the transmitter off the air and in some cases starts the cell ionization. It is for this reason that the cards bearing the station's call letters are printed in white on a black background. Such cards facilitate control of the signal, but nevertheless, precaution must be taken by the operator focusing the program to lower the focus in time to overcome quick changes in volume, much the same as in sound broadcasting.

ADAPTING FILMS TO TELEVISION.—The motion picture theater is said to be right at the front door as soon as television lifts up the latch and walks in.

Apparatus is being developed so that the standard films can be projected by television. One of the greatest difficulties encountered in this adaptation of the sound-sight film to television was caused by the difference in the rate of speed with which the pictures are taken on the movie lot, and that with which they are scanned in the television studio. The motion picture camera exposes twenty-four sections of the film each second, whereas television laboratories have experimented with scanning systems projecting a maximum of twenty pictures a second.

The consequent slowing down of the film results in slow motion of the characters in the television picture. Furthermore, the slower movement of the sound track past the photoelectric cell creates a sound distortion such as that noticed when the turntable of a phonograph revolves slower than the speed at which the recording was made.

Armando Conto, research engineer of the Western Television Corporation, has developed apparatus that broadcasts the standard film with the characters moving at normal speed, and the sound taken undistorted from the reel. He uses a three-spiral scanning system which divides any area to be broadcast into forty-five horizontal parts at a speed of fifteen times a second. This leaves a considerable gap between the fifteen pictures a second as broadcast by the television station and the twenty-four pictures a second projected in the cinema theaters.

Conto looked with disfavor upon the method by which the film is kept in motion as a part of the scanning operation, a practice used in previous technique. He decided that superior results could be obtained if the film remained stationary, as it does in the projection of motion pictures, moving forward at a predetermined speed. So he designed a disk that combines the effects produced by an ordinary scanning device and the shutter on a moving picture projector. The disk is built so that the holes through which the light penetrates are placed on radii four degrees apart instead of eight degrees as in the ordinary three-spiral 45-hole disk. Thus, the forty-five holes occupy a 180-degree segment of the disk, leaving the other half blank to act as the shutter. Two identical films are employed. One reel is located at the upper diameter of the scanning disk and the other at the lower. The movement of both films is toward the center of the disk.

The films are placed in the same position in each Geneva movement (the device which moves the films forward). The two Geneva movements and the scanning disk are interconnected mechanically in such a way that when the first hole of the spiral is in a position to scan picture No. 1 in film No. 1, this film remains motionless for the duration of the entire scanning operation. While this is being done the blank segment of the disk is passing before film No. 2, shutting off the light. This interval is used to move picture No. 2 in film No. 2 to a standstill position so that it will be scanned immediately after picture No. 1 of reel No. 1 is scanned.

Several experimenters are trying the scanning disk with the holes arranged in a circle instead of spirally, as a method of utilizing the standard sound-sight films in television projection. When this type of disk is used the film moves steadily with no intermittent motion, whereas with the spiral hole arrangement the film does not run smoothly but with an intermittent motion.

Is a New Name Needed?—There has been some discussion relative to a name for television set owners. Listeners is a logical cognomen for those who tune in on sound broadcasts.

Alexanderson has suggested the name "radio spectator" to apply to the owner of a television set. The receiver, he believes, might be called a "teleopticon," but he hopes that no such linguistic abomination as "televisor" will be used. Aylesworth thinks "radio audience" is superior to any newly coined word. "Spectauditor" is suggested by George B. Cutten, President of Colgate University.

Frank P. Day, president of Union College does not see how a new word can be coined for a television receiver any more than for an ice box or kitchen stove. The obvious word for the user of a televison set, however, might be "televist." DeForest offers "televiewer" and "teleseer." John Grier Hibben, president of Princeton believes "observer" might be satisfactory because observation is the function of both eye and ear. Harold LaFount, Federal Radio Commissioner, agrees with Hibben, because "observer" is all-embracing and in no sense misleading. Dr. Michael I. Pupin presents "televisioner."

"I generally prefer straightforward, blunt, Anglo-Saxon terms," said Dr. Alfred N. Goldsmith. "Tortured Græco-Roman terms, evolved by ingenious lexicographers in cloistered studios rarely appeal to the public. When we want a man to watch what is happening at a railroad crossing do we say: 'Decelerate, Observe Visually; and Ausculate'? What we say is 'Stop, Look and Listen!'

"The public, with good sense, has decided that we 'listen-in' to radio programs, and has called itself a group of 'listeners.' Likewise the public will 'look-on' television pictures and will probably be willing to be called a group of 'lookers.' But when it comes to those who both look and listen, the problem is more complicated. Therefore, I suggest the coined word 'lookstener' which is a sort of abbreviation of look-and-listener."

Many other words have been proposed such as viseur and looker-in, but "observer" seems to have the best chance for being generally adopted.

Chapter Twelve

TINY WAVES THAT "SEE"

It begins to look as if television's destiny is bound up in little radio waves—waves that must be projected from on high because in general they act almost like rays of sunshine. Television is leading man into unfathomed realms of science, into a spectrum long considered useless, but nevertheless mysterious.

It is helpful in trying to comprehend the possibilities in the short-wave realm to refer to light, and to remember that the eye perceives light waves as long as 40,000 to the inch or as short as 80,000 to the inch. All the colors between red and violet, and all the scenes that man beholds fall within this microscopic range. The longest red is about one-sixty thousandth of an inch longer than the shortest visible violet wave, according to Henry D. Hubbard of the Bureau of Standards. It is believed that outside this range there is a vast spectrum of unseen rays. Visible sunlight carries only one-fifth of the sun's total energy. Scientists are seeking new "eyes" that will perceive some of these unseen light waves.

Today there is an infra-red camera that takes pictures in the black of night. The photographic plate is no longer color blind to infra-red that contains 80 per cent of the sun's radiant energy. And so in radio and television new and startling possibilities lurk in the ultra-short waves. It is the greatest field in radio research today. The television camera may eventually utilize infra-red rays that will empower it to see what is going on at night. Radio is just scratching the surface of this spectrum which promises unique developments in the wizardry of television. The images may even travel on a beam of light!

TINY WAVES THAT "SEE"

Micro-Rays Surprise the Experts—May 1, 1931

Amateur experimenters everywhere, and a corps of professional engineers steeped in the lore of research, are creating the little waves, flinging them into space and then recapturing them to see what they do in the air, and what they can achieve in transmission of sound and sight.

Technically, they are called ultra-high frequencies or micro-rays. The layman refers to them as tiny waves, and that, too, is correct. Physicists have named them quasi-optical, because of their close relationship to light.

Achievements already credited to this spectrum indicate that it may hold the key to more than one ethereal lock which will unfetter television images and free them from their scientific prison. It is true that these short waves act in a freak, uncanny manner at times, but they call for simple, inexpensive, compact apparatus and comparatively low power. They are economical. And last but by no means least, ultra-short wave stations can be packed almost as close as sardines in a can. There are limitations in the distance they cover, but in the limitation of radiation there is found an important asset.

Already, waves measuring only seven inches from crest to crest, are carrying voices across the English Channel between the cliffs at St. Margaret's Bay, Dover, England, and Blanc Nez, near Calais, France. As a result, novel possibilities for television progress are foreseen. The new system is heralded as a revolutionary development.

Engineers of the International Telephone and Telegraph Company, employing a miniature station equipped with an antenna just an inch long, radiating power estimated at half a watt, enough to operate a small flashlight bulb, have triumphed across a stretch which marked one of Marconi's early achievements in wireless communication.

It was on March 27, 1899, that the inventor of wireless signaled from Dover to Boulogne. At 5 o'clock in the afternoon Marconi pressed the key releasing the sparks for a jump of thirty-two miles across the Channel. That was a long distance for wireless in those days. Records do not mention the length of the wave used in the 1899 experiment. It was probably about 150 meters, because the possibilities of short waves were undiscovered.

Since that day, however, both amateur and professional experimenters have learned many of the secrets lurking in the elusive short waves. Boys have talked around the world using less power than is required to operate their mother's electric iron.

EARTH'S CURVATURE INTERFERES.—The new system that spans the Channel is called micro-ray radio. The results are said to have surpassed the most sanguine expectations of the engineers. They recall that only two years elapsed after Marconi's Channel success before he picked up the first transatlantic signal. The engineers at the laboratories at Hendon (England) and Paris are planning further refinements and new developments, which they hope will make possible everyday commercial applications. Eventually, they may find a way to send these waves across the ocean, but today the curvature of the earth stands in the way, unless a number of ocean-relay stations were used and that is impractical.

A remarkable fact is that the tiny waves do not fade. They carry the voice clearly. An ingenious combination of two reflectors concentrates the radio power into fine pencil-like rays and projects them into space in much the same manner that a searchlight casts a beam of light. The reflector is about ten feet in diameter. It faces the direction of the distant receiver. Another set of reflectors intercepts the radio beams.

Where there is a sending and a receiving station on the same site, for example, at Blanc Nez, the receiving outfit is built eighty yards from the transmitter and arranged to

TINY WAVES THAT "SEE"

be in its electro-optical shadow, adequate allowance being made for diffraction. The same wave length is used for both transmission and reception.

"The success of the demonstration has definitely shown that wave length range as low as 10 centimeters is opened up," said Frank C. Page, vice president of the International Telephone and Telegraph Company. "The importance of this from the point of view of relieving radio congestion need hardly be stressed. A simple calculation will show that the range of frequencies available within this band is some nine times as great as that in the wave band heretofore used. Added to this is the fact that the radiations can easily and cheaply be concentrated into a small, single band or conical ray. The frequency band now available will permit the working of a large number of permanent and continuous channels between the same places without mutual interference, while the directional properties and comparatively short range of the waves will make possible the use of the same frequencies or waves for other routes.

NEW HOPE FOR TELEVISION.—"A further important use will be for television transmission," said Page. "The present difficulty with regard to television is the large frequency range (wide path in space) required for satisfactory definition of the object that is broadcast. It should be possible to allocate as wide a band as is necessary for television without causing any other congestion. It is easy to imagine the establishment of national micro-ray networks for use in conjunction with television apparatus.

"For navigation purposes and especially for radio beacons, the simplicity of the transmitters has obvious advantages. Valuable applications seem possible in ship-to-ship communication, as the small size of the equipment would enable easy use of its directional properties. This, coupled with the short range, affords a satisfactory method for virtually secret intercommunication between war vessels."

A radio wave 25.70 inches long carrying a signal clearly over sixteen miles reveals to the amateur that he is by no means alone in finding that electromagnetic channels below 10 meters have a wide field of usefulness, but not for long distance. Indications point to the probability of a future ultra-high frequency spectrum swarming with telephone communication systems, television transmitters and broadcasting stations.

Kenneth B. Warner, an official of the American Radio Relay League, contends in the magazine *QST* that somewhere around 43,000 kilocycles (7 meters) is a limiting frequency the sky waves of which seem never to return to earth. That wave is believed to mark the upper limit of frequencies useful in the ordinary methods of transmission. Although the frequencies up to that value are useful for long distance operation, including at times the 10-meter band, the frequencies from that point up seem valuable only for short distances.

"From the meager literature, one judges that the receiver should be able to 'see' the transmitter," said Warner. "If a hill intervenes, the transmission is likely to be cut off. Curvature of the earth limits the range to the distance where the wave becomes tangent to the earth; therefore, the higher the transmitter the greater the range. Limited application? Not at all. This is just the thing for commercial television (if a satisfactory technique is developed), because here is unoccupied territory sufficient to accommodate the enormous modulation bands required and beautifully limited in range. The very peculiarities of these frequencies, in that they cover limited mileage, enable television stations to duplicate their use on the same wave length in every city in the land without interference.

High Aerial Is an Asset.—"We understand that commercial television developments looking to these ends are in process. Of course, the aerial will have to be on a high mast,

TINY WAVES THAT "SEE" 181

or located on a hilltop, or perhaps even suspended over its 'service area' by a small balloon—but those things will work out. Aviation finds these frequencies of even greater promise, and for similar reasons. In Hawaii the public telephone service on the various islands is interconnected by short-wave radio links a few meters long, the stations being located on mountaintops to 'see' each other and to clear the curvature of the earth between. In such transmission there is no fading, no static, no uncontrollable interference from other stations. These tiny waves create 'a radio heaven' for short ranges."

WITHIN A SMALL ORBIT.—Down to perhaps a meter or two, the experimenters have discovered that more or less ordinary circuit arrangements can be applied, according to Warner's report. He explains that below that, in the region of centimeters, a most fascinating world awaits the experimenter with Barkhausen-Kurz oscillations, the frequency of which is not determined by inductance and capacity but by the orbits which electrons trace inside the tube, where wave length is controlled not by tuning but by the varying of voltages. And these oscillations are apparently rather easily produced in appreciable power.

A Japanese experimenter has reported to the Institute of Radio Engineers that he sent signals over several miles on a wave length of less than half a meter. It has been observed that the extremely short waves do adhere to the same laws, apparently, as those of a few meters in length, but they call for short aerials, and reflectors the dimensions of which make them easier to handle.

In this connection some years ago John Reinartz, one of Connecticut's noted radio amateurs, was experimenting with short waves and found that the copper bowl of an ordinary household electric heater provided an excellent reflector.

"Doesn't that excite your imagination and cause a few day-dreams as you visualize reflectors and beam systems in miniature, quickly built, easily changed?" Warner asked the amateurs. "We should find it interesting to participate in the development of the frequencies above 56 megacycles, particularly in the creation of apparatus that will work well in these regions. It is a rich, new field, fertile with possibilities for the ingenious, and undoubtedly destined ultimately to have a big part in amateur radio. Just what that part is, our experimenters will determine."

WHAT ENGINEERS OBSERVE.—The applications of frequencies above 30,000 kilocycles were discussed before the Boston section of the Institute of Radio Engineers in a paper prepared by H. H. Beverage, H. O. Peterson and C. W. Hansell, engineers of the Radio Corporation of America. Their observations are based on experiments over a period of several years.

They find that the altitude of the terminal equipment location has a marked effect on the signal intensity, even beyond the optical range. Frequencies below about 43,000 kilocycles appear to be reflected back to earth at relatively great distances in the daytime in north-south directions, but east-west transmission over long distances is extremely erratic.

Frequencies above approximately 43,000 kilocycles do not appear to return to earth beyond the ground-wave range, except at rare intervals, and then for only a few seconds or a few minutes. Ground waves which are not bothered by a sky wave returning to mingle with them also appear to be free of echoes and multiple path transmission effects. Therefore, they are free from distortion due to selective fading and echoes. The range is also limited to the ground-wave range, so these frequencies may be duplicated at many points without interference. For example, stations in New York, Providence and Philadelphia could use the same waves without overlapping. Experiments with frequencies above 300,000 kilocycles have so far indicated that the maximum range is limited to the optical distance.

BUILDINGS ACT AS REFLECTORS.—Tests are under way to determine the shielding effect of city buildings. A receiver is mounted in a test car and continuous observations of a 60,000 kilocycle signal from Weehawken are conducted while driving through various streets on the Manhattan side of the Hudson River. The transmitting aerial is on the roof about 100 feet above the river. It is found that the signal can be heard in the streets four or five city blocks back from the water front. Driving along a street parallel to, and several blocks back from, the river, it is noticed that the signal strength increases greatly whenever one of the streets perpendicular to the river is crossed. This suggests that buildings may serve as fairly effective reflectors. Thus, it may be possible to obtain quite effective broadcast service to all parts of a city if the transmitter is atop a tall building. Because of obstructions in city areas, however, it is believed that the service range will be restricted for high-grade entertainment broadcasting and television. High power is an essential factor in overcoming some of the ill effects.

VARIOUS USES ARE OUTLINED.—In conclusion the engineers enumerate some of the uses for which the short waves might be especially well suited. They explain that frequencies above 43,000 kilocycles do not seem to be reflected back to earth by the Kennelly-Heaviside surface. They fly off tangent from the earth like sparks from a grinding wheel. And, furthermore, as the frequency is increased, the maximum range tends to approach the optical range as a limit.

This is a fortunate limitation and in many cases should be advantageous. It should eliminate the fading effects and distortion so troublesome on the lower frequencies. It should be possible to use the same waves over and over at geographically separated points on the earth. A few ultra-high frequency applications are outlined by the engineers as follows:

(1) Point-to-point communication up to 300 miles between mountains.

(2) Ground-to-aircraft communication up to at least 100 miles and communication between aircraft.

(3) Point-to-point communication between high buildings or towers up to fifty miles or more.

(4) City police alarm distribution up to a few miles with portable receivers carried by patrol cars.

(5) Possible application to high speed visual image distribution over local areas.

(6) Local audio, facsimile or ticker distribution.

(7) Communication and direction finding for ferryboats, tugs and harbor craft.

(8) Marker beacons for air and water craft.

AERIALS GIVE NOVEL RESULTS.—Harold H. Beverage and Dr. N. E. Lindenblad are conducting ultra-short wave experiments at Radio Central on Long Island. They have linked Rocky Point and Riverhead, fifteen miles apart, by little waves radiated by aerials thirteen inches long held aloft thirty feet to dodge the earth's curve.

They sent an airplane up with a receiver on board. Another set was located atop the Empire State Building. The operator could not hear the message on the 68-centimeter wave (27 inches), but the airplane picked it up at an altitude of 1,600 feet. The beam was only a mile wide. When the plane flew out of that range the signals vanished. Then they put the aerial up on an 80-foot mast at the transmitter and the receiver on the skyscraper detected it too.

The engineers are not positive how the 68-centimeter waves travel. They are reasonably sure that in general they follow the curvature of the earth. On several occasions, however, it has been possible to detect signals several miles beyond the horizon point. It is believed that moisture in the atmosphere reflects the signals to a slight extent, making it possible for them to bend a few miles below the horizon.

TINY WAVES THAT "SEE"

So far only a few watts have been carried by these waves. The engineers want them to carry more energy. They are trying to find out the greatest amount of power that can be handled on the shortest possible wave. The tubes are a limiting factor. New ones are being designed which may make it possible to use several kilowatts on ultra-short wave channels.

They have observed that the type of aerial employed is important. One arrangement of the sending wire radiates 68-centimeter waves over an area shaped like a huge doughnut. The transmitter occupies the exact center of the hole. An aerial of different design covers an area in the form of a large slice of pie. An aerial of megaphone shape, with the transmitter as the mouthpiece, sends out a slender beam of energy. The aerials are mounted on a board about six feet square faced with copper and surrounded by reflector aerials. Connection between the aerial, atop a high tower, and the oscillator is made by means of radio frequency feeder lines, which because of their design merely carry the energy from the transmitter to the aerial or radiator where it is reflected in the desired direction.

NEW TYPE OF INSULATOR.—A unique insulator, which in reality is an electrical conductor designed to function as an insulator of the radio frequency current, has been developed for the aerial system of the 68-centimeter oscillator. It consists of two wires each equal to one quarter of the wave length, or 17 centimeters long, arranged in parallel and linked to a ground connection at one end. If a wire equal to one-half the wave length, or 34 centimeters long, were stretched out in a straight line, it would form an aerial identical to the one located on top of the mast. By running the two wires parallel, about two inches apart a simple tuned circuit is formed. It has inductance and capacity. The radio frequency field of the wires is concentrated within the center of the two wires because of the parallel arrangement, and therefore, has zero radiation resistance. It is really a hairpin of wire forming a tuned circuit that traps any energy from leaking away. This forms a highly efficient high frequency insulator and, according to Lindenblad, is more efficient than glass or porcelain insulators. The insulator may be used on wave lengths up to five meters.

While millions of cycles are involved in broadcasting at such high frequencies, practically no wire and condenser capacities as found in the usual radio transmitter or receiving set are utilized in this apparatus. Connections between the various parts consist of straight wires which form the tuning inductances and capacities. The circuits are composed of numerous copper rods, which may be lengthened or shortened by sliding in and out of one another.

THE BARKHAUSEN TUBE.—The standard vacuum tube does not act well when called upon to perform in the ultrashort wave realm. It refuses to oscillate at high frequencies, that is, below two meters. The electrons inside the bulb do not travel fast enough to support the exceedingly rapid electrical vibrations required. So a new tube has been built ingeniously to speed up the electronic action and it operates easily at high frequencies. Dr. Heinrich Barkhausen of Germany developed the tube and observed the result, now called the "Barkhausen effect."

The technical aspects of this tube, as discussed in *Electronics*, calls attention to the fact that in generating ultra-high frequencies the self-supporting grid of the Barkhausen oscillator is maintained positive. The plate is negative. Then electrons emitted by the cathode are attracted to the grid. Many of them pass through the grid's mesh and get within the field of the plate. Inasmuch as the plate is negative, the electrons are repelled and thrown back to the grid. One oscillation, therefore, takes place in the time required for an electron to make its circuit. For successful operation the plate might have a negative voltage of 40 while the grid has a positive voltage of about 250 placed on it.

Voltage is applied to the tube through choke coils to protect the power supply apparatus from the high frequency currents and to prevent part of these supply leads from forming an oscillating system at a frequency lower than that desired for radiation from the aerial proper. Since the wave length is a function only of the size and the voltages thereon, the tube is designed in such a way that all parts serving as coupling devices are exact ratios in size and spacing of the desired wave length. A shield about one inch square protects the radiating parts of the bulb from the field of the aerial. The bipoles going out from the grid and plate are carried to the focus point of the reflectors. At the receiving end wires on a frame about two feet long act as an antenna to collect the radiation whence it is conducted to a tube similar to the one used for transmission except that it is designed for lower voltages.

CHARACTERISTICS OF THE WAVES.—Obviously, the outstanding characteristic of the ultra-short waves is straight-line propagation. Between two points there is always only one line of propagation and for that reason all the phenomena of fading are unknown at quasi-optical waves. Another important feature is the possibility of concentrating energy. Furthermore, the noise level is extremely low. E. Karplus, engineer of the General Radio Company, has observed that this seems to be due to the fact that even nature has some difficulty in starting these high frequencies and that they do not occur in man-made devices such as electric signs, elevators, motors and lamps which at times interfere with broadcast reception.

It has been proved by theory and practice that for the longer waves, that is, down to 5 centimeters, humidity, rain and fog have no influence on propagation. Below 5 centimeters, however, the engineers notice an absorbing effect caused by humidity and especially the content of carbon dioxide. Waves below 3 centimeters have no appreciable radiation in the atmosphere. They are absorbed and scattered in the immediate vicinity of the transmitter. Radiation of electromagnetic waves that permit communication begins again only at the shorter heat waves, at the infra-red and light range. Attenuation of these waves is somewhat less than in the range of visible light.

"So far as modulation is concerned," said Karplus, "quasi-optical waves are much better off than all other waves used in communication. This may be of great importance when all the difficulties that limit television today are eliminated. Ten meters has been assumed arbitrarily as the upper limit of quasi-optical waves. It is impossible, of course, to draw distinct limits in nature and it would probably be better to say 5 meters instead of 10, but the choice of 10 meters was dictated by the fact that waves below 10 meters only occasionally are reflected back from the upper atmosphere. Short-wave broadcasting experiments have been conducted in Berlin. After tests at 3 meters, the wave length was shifted to 7 meters. With a one kilowatt transmitter located on the roof of a building about 100 feet high satisfactory results have been attained up to a distance of five miles."

OLD IDEA COMES IN HANDY.—When waves below seven meters are used the super-regenerative circuit developed by E. H. Armstrong back in 1922 finds a new field of usefulness. Its stability, simplicity and high amplification adapt it to reception of pictures. It will be recalled that the regenerative circuit was extremely popular in the early days of broadcasting, but it howled and squealed so much that it was frowned upon. When tuning it, the sensitivity builds up rapidly with regeneration until a point is reached where it oscillates. That chokes the tube and the signals disappear.

Armstrong sought a way to avert the choking and still

retain the benefits of regeneration, and in so doing he developed the super-regenerator.

ENTERING THE PROMISED LAND.—"Any engineer who has been confronted with the problem of allocating channels to all the manifold demands for radio service must quickly become impressed with the fact that a 'ceiling' has always overhung the usefulness of space radio," remarked O. H. Caldwell, former Radio Commissioner. "Nature has provided only one ether spectrum, and all classes of radio service must accommodate themselves to its sharp limitations. Multiplication of the spectrum, it was apparent, could come only through subdivision or else by expansion into the higher frequencies.

"Recent work with television in the short waves and with quasi-optical waves, now seems to open up a vast new realm for radio service. Kilocycles have always been the crying need in radio, and as we go down into the short waves, we turn up kilocycles in profusion. For, every time we halve the wave length made useful for radio, we add to the former spectrum as many kilocycles as we had, altogether, before!"

Range	Added Channels	Total Available Spectrum
Infinity to 10 meters (present spectrum)	....	30,000 kc.
10 meters to 5 meters	30,000 kc.	60,000 kc.
5 meters to 2½ meters	60,000 kc.	120,000 kc.
2½ meters to 1¼ meters	120,000 kc.	240,000 kc.
1¼ meters to ⅝ meter	240,000 kc.	480,000 kc.
⅝ meter to 5/16 meter	480,000 kc.	960,000 kc.
5/16 meter to 5/32 meter (15.6 cm.)	960,000 kc.	1,920,000 kc.

At last, there are kilocycles enough, and down among the short waves the same frequency can be used over and over again, even in the same locality. Thus the lid is lifted. The limits on the multiplied use of space radio become only those of equipment and demand.

It is possible that private-line telephones, mechanical control and a host of other uses may follow, until the cities and countryside of the future are everywhere cross-threaded with "wireless" local circuits. Caldwell says that with simple terminal sets and the cost of intervening wires eliminated, even a vivid imagination wonders what may be the uses of this new radio realm. Here may develop an equipment market that will parallel broadcast receivers in numbers. Engineers and manufacturers will do well to watch this quasi-optical field closely.

Results on the frequency band from 43,000 to 45,000 kilocycles have so enthused Boston experimenters at station W1XG, that they call the 6.97 meter wave "the radio man's paradise." The tests so far indicate that the service area of this channel is about forty miles.

"Where a single broadcasting station requires a niche only 10 kilocycles wide a television station needs 100 kilocycles," said Hollis Baird. "In the broadcast spectrum from 550 to 1,500 kilocycles there is space for about ninety-six cleared channel stations that send out voice and music. Ten television transmitters would entirely fill this space and be uncomfortably close to each other. One can easily figure out that in the ultra-short wave spectrum, let us say, from 30,000 to 100,000 kilocycles (10 to 3.3 meters), there would be room enough for 7,000 broadcast transmitters and 700 television stations, all on cleared channels. No fading, no static! No wonder we are excited over the possibilities of these waves."

A good idea of how these waves dodge static is found in a report from Hawaii where the inter-island radiophone makes use of the ultra-short wave channels: "On Friday and Saturday we passed through one of the greatest static storms in my experience here on the island," said an engineer. "Lightning flashed almost continually for the better part of two days and nights. No difficulty was experienced, however, in the operation of our ultra-frequency telephone circuits. It was a weird experience to be watching the lightning and at the same time talk with Hilo without difficulty or without any particular annoyance being experienced from the faint indications of static on the circuit."

Chapter Thirteen

A FLYING SPOT OF MAGIC

Television's flying spot of light is thrilling the world. A switch is thrown. A motor purrs. Electrical life is instilled in the copper arteries and veins. The electric nerves tingle and a spot of light flashes on the screen. It moves slowly at first, then gains in speed as other spots appear. They all move fast, gyrate and streak the screen. Soon it is flooded with light. Out of it all comes an image!

So Near and Yet So Far—June 1, 1931

How near is television to the home? Discussion of that subject is usually prefaced with the general statement that it is just around the corner. No one seems to have discovered what corner; whether it is where Zworykin Avenue crosses Ives Street, where Baird Avenue meets Alexanderson Boulevard, or where Farnsworth Road crosses Sanabria Lane.

Progress is being made but the images reaching the home today on tiny "screens" are not of sufficient quality or size to engross the family attention for any length of time as broadcasting does. Nevertheless, the television era has definitely dawned, according to the observations of Aylesworth. And Sarnoff asserts that transmission of sight by radio is a matter of accomplishment, not of speculation. He believes that the present sporadic activities cannot be classed as practical service. They are purely experimental, but as such deserve encouragement and merit public interest. He likens the present status of television to the pre-broadcasting era of radio, when amateur experimenters were beginning to hear faint sounds in their earphones.

THE EXISTING PROBLEMS.—"The next stage in television —and I should anticipate its realization by the end of 1932 —should find it comparable to the earphone days of broadcasting," said Sarnoff. "At this point the public may well be invited to share its further unfolding. By that time, television should attain the same degree of development as did sound broadcasting in the early period of the crystal set. In the practical sense of the term, television must develop to the stage where stations can broadcast regularly visual objects in the studio, or scenes occurring at other places through remote control; where reception devices shall be developed that will make these objects and scenes clearly discernible in millions of homes; where such devices can be built upon a principle that will eliminate rotary scanning disks, delicate hand controls and other movable parts; and where research has made possible the utilization of wave lengths for sight transmission that will not interfere with the use of the already overcrowded channels in space.

"Important forward strides are being made. In our development laboratory at Camden we are seeking to perfect television to a point where it is capable of rendering real service. While the public was willing, and even eager, to experiment with radio in the early stages of broadcast development, it seems to us that it will desire a comparatively more advanced television receiver than the early crystal radios. There was no precedent for the taking of sound and music out of space, but the public has been educated by the motion picture industry to expect picture transmission of a high quality, and it is doubtful whether interest can long be sustained by inferior television images.

"The progress we have made so far has given us the belief that ultimately a great service of television can and will be made available. I do not believe that television will supersede sound broadcasting. It will be a correlated industry. Television promises another great industrial development, but to assure this, we cannot disappoint the public and defeat the possibilities of a future great service by hasty and premature action at the present time.

"Last year I said that perfected television would come within five years. The results of our work in the past six months has brought the goal some years nearer."

Further inquiry among leaders in the radio field reveals a diversity of opinion regarding television's possibilities.

"I believe television will be in operation on a commercial basis by the end of 1932," said William S. Paley, president of the Columbia Broadcasting System. "However, people should not expect too much. There is a great deal of pioneering and experimenting to be done. One of the big jobs identified with the coming of television, in addition to the technical and production development, will be the reorganization of broadcasting to conform with the new requirements of sound and sight."

DARKNESS NOT DESIRED.—"Television is in the home right now!" exclaims Clem F. Wade, president of the Western Television Corporation. He points to the fact that 3,500 visual receivers are in the Chicago area.

"Pictures received in homes have been small," said Wade. "A darkened room has been necessary on account of the feeble illumination. This has limited the sale and use of the set. We believe that television will receive the same impetus that the loudspeaker gave to radio when a larger picture is shown in the home without darkening the room. It will not be long before a picture six inches square will have sufficient illumination to be seen in daylight. In darkness, the size may be increased to several feet square."

AGITATION IS PREMATURE.—Harold A. Lafount, Federal Radio Commissioner, finds it difficult to predict how long it will take to perfect and commercialize television. He foresees many perplexing obstacles, which must first be overcome before one can state that television is in the home. Lafount believes that three years is an optimistic estimate.

"In my opinion," said the Commissioner, "the present agitation and interest in television are premature and may give the public a false impression. It would be a severe blow to the radio 'infant' to call upon it at this time to do a man's job."

ON WINGS OF PROSPERITY.—Dr. Lee de Forest asserts that we are perhaps nearer to television in the theater and further from television in the home than the majority of people realize.

"With the return of general prosperity there is no question that radio manufacturers will intensify their efforts to revive, by way of wholesale television manufacture, their 'old-time' prosperity," said de Forest. "The industry seems a unit in the conviction that nothing but television can really restore this; and under the spur of the lash, improvement in home television technique may surprise many who are today pessimistically inclined."

RESULTS CALLED CRUDE.—Powel Crosley, president of Crosley Radio, reports that he and his engineers have watched and studied everything they can find in television, but so far "we have seen nothing that belongs any place except in the laboratory."

"In the last twenty years only comparatively slight improvement has been made—slightly better photoelectric cells, slightly better illumination for the picture," said Crosley. "We feel that it is not time yet to get the public worked up over the present crude results. The scanning disk seems to limit television to an interesting laboratory experiment. The lack of broadcasting channels and the necessity for wide frequency bands required to make reasonably good pictures seems at this time to bump it into an almost impossible situation."

A Scientific Novelty.—Ray H. Manson, chief engineer of Stromberg-Carlson Telephone Mfg. Co., contends that television is a scientific novelty of great promise, and so long as the public is not led to expect too much from the systems now in use, progress can be made in an orderly, satisfactory manner.

"Larger pictures with more detail and better fidelity are necessary before television can be considered commercial," said Manson. "Also, the pictures must be so arranged that fairly large groups of observers can look at one time. It is reasonable to expect that any great stride in the advancement of television will be through some new invention for simplifying the transmission problem. Otherwise, progress will be comparatively slow, and the public will have to wait several years for the commercial results."

"Now that we have television, what shall we do with it?" asks Hollis Baird. He answers the question himself:

"One of television's first steps will be the projection of talking picture films, which will bring to the home entertainment featuring sound and sight. This is the result of years of work by the motion picture producers. In addition mere news flashes need no longer be broadcast audibly. News events recorded by sight and sound can be put on the air the day they happen, in the evening when many will be at home to enjoy them.

"Then comes the more involved question of studio productions or direct pick-up entertainment. New photo-cell equipment permits close-ups and long shots so that television has variety which was lacking at first. Fading-in from one of these 'shots' to another can be accomplished electrically as easily as a motion picture fades from one scene to another. This brings up the question of scenery. How much background can be picked up? That will depend on the scenic effects. Undoubtedly suggestion and exaggerated details will make up the earliest scenery. And if make-up can

TALKING BY TELEVISION

Dr. Frank B. Jewett, president of the Bell Telephone Laboratories, with Dr. Frank Gray in the television-phone booth.

FOUR POWERFUL EYES

Station W2XBS atop the New Amsterdam Theater, installed to experiment with television in the New York area.

A CLUSTER OF SEVEN LIGHT SPOTS

The drum of mirrors begins to whirl. The dots of light start to gyrate, and soon the entire screen is flooded with light. Then an image appears. Alexanderson points to the magic cluster.

help a motion picture actor with the fine definition which the movies permit, it will surely have a big place in television.

"Simple variety or vaudeville acts lend themselves easily to television, but the dramatic field has richer possibilities. The popularity of radio dramatic skits proves that the public enjoys this type of entertainment despite the limitations of acting that comes to the ear only. There are wonderful possibilities in television drama."

Two Camps Are Found.—Ross A. Hull, associate editor of *QST*, has made a survey of television for the benefit of radio amateurs, and he finds television interests divided into two camps: those anxious to talk and those anxious to avoid talking. The most voluble unfortunately have the least information on the subject. The non-talkers have crawled into their shells to avoid playing a part in the premature and misleading publicity. Then, too, they have inventions to protect.

A summary of Hull's observations reveals: Sixty-line pictures provide a momentary thrill—they fail to keep the family at home engrossed in a television program . . . the cathode ray tube has been shown to promise an effective way of scanning. It has every indication of being one logical successor to the scanning disk, free from the inaccuracies, the inconveniences and the speed limitations of any mechanical device. . . . It is not certain that ultra-high frequencies are capable of good service. . . . Wire linkage of stations throughout the country probably will still be impractical because of limitations of wires in carrying high frequency currents . . . with 240 lines to a picture there will be little danger of mistaking the soprano for her poodle . . . there is a big fire in the television stove but the cooks are still without a recipe book. . . . Television of the moment is an intriguing and utterly absorbing field for the experimenter but as entertainment it is still around the corner.

The Gap Is Closing Up—June 2, 1931

The gap between those who believe that television is already here and those who concede that it is still around the mythical corner is steadily closing up. A television station is being erected at a cost of $85,000 atop the Empire State Building, the world's loftiest skyscraper. It is expected to be the engineers' most practical teacher, and it may stir up an interest and a curiosity among the public to look in on what is passing through the New York air.

Horizon Is Variable.—The technical horizon from the observation tower of the building is sixty miles. However, there is usually a haze that hides everything beyond the thirty-mile limit. When places forty miles distant are recognized it is an exceptionally clear day. Twenty-five miles is considered to be a good average range. It is reported that Patchogue, Long Island, has been seen to the east on a clear day, Ossining to the north, the Orange Mountains to the west and the open sea to the south.

If the experimenters find that the technical horizon is the absolute limit that the quasi-optical waves will cover, then a sixty-mile radius with the Empire State tower as the center will be the range of that station. However, the experts will be gratified if they can serve that area with a single transmitter, because within that circle lies the most thickly populated land in the country. Such a television might claim a vast audience should homes be equipped with vision sets as they are with broadcast receivers. New York's metropolitan area, according to the Census Bureau's 1930 figures, has a population of 10,901,424, and within 2,541 square miles. A high power station might reach them all.

The power is rated at 5,000 watts, and it will filter through space on the following channels by authority of the Federal Radio Commission: 43,000 to 46,000 kilocycles; 48,500 to 50,300 kilocycles, and 60,000 to 80,000 kilocycles.

A FLYING SPOT OF MAGIC

Another transmitter of 2,500-watt capacity will be utilized for experimental purposes. It is licensed to use 41,000 to 51,000 kilocycles; 60,000 to 400,000 and above 401,000 kilocycles.

The station is located on the 86th floor, 1,000 feet above pedestrians in the street. The aerial is a fourteen-foot rod on top of the mooring mast, which makes the pinnacle 1,276 feet high. The engineers are hopeful that operation of the visual broadcaster at this altitude will be helpful in surmounting the difficulties that beset television transmission in the city. If they can succeed in New York with its massive steel structures, then television in other cities will be an easy task. This lofty station is designed to get the images well on their way before the steel fingers have a chance to clutch them, and the high power is depended upon to drive the faces through "dead spots" or so-called shadows, which the buildings cast in the path of radio.

While the research experts have done exceedingly well with the crude, basic television principles, they consider the commonplace technique comparable with attempting to design a wrist watch with locomotive parts. The first means in a new science are obviously immature and cumbersome with regard to the delicate end. Today nothing can compare in simplicity, low cost and practicability with mechanical scanning; therefore, the only course lies in further refinement and improvement of components and assembly or the discovery of a more efficient system.

LIMITATIONS ARE A CHALLENGE.—"The present limitations imposed on television are no greater than those imposed on early broadcasting," Aylesworth once remarked. "It has not always been possible to broadcast an entire symphony orchestra with every assurance that the reproduction would be successful. In the early days large orchestras were avoided by broadcasters with reputations to maintain. Instead, a few musicians were selected. To go beyond a few musical instruments was to court disaster. Those who attempted complete orchestras presented their audience with a radio version of the Tower of Babel."

Those who have watched a clean-shaven tenor appear on the television screen with a goatee, or Mayor Walker of New York appear with a mustache, realize that existing radiovision instruments have limitations. Only a small amount of detail is available. With just so many light elements at hand with which to assemble the images at the receiving station, it is necessary to work with large figures or close-ups, or to sacrifice detail in obtaining a larger field of vision. It is possible, therefore, to reproduce close-ups of personalities, with facial features discernible, so that identification imposes no severe strain on the imagination. Half-length pictures result in marked loss of detail. Facial features are insufficiently distinct to permit quick identification. However, a greater range of action may make up for loss of detail. Full-length pictures or so-called long shots possess little detail. Action alone must tell the story because the figures may be virtually silhouettes.

The limitations of television are simply a challenge to the ingenuity of the broadcasters, as Aylesworth sees it. He believes that the program presentations can in large measure be fitted to the limitations, even giving birth to a unique form of art, perhaps, as in the case of the silent motion pictures and sightless broadcasting. Television is more fortunate in its early struggles than was sound broadcasting, because while the latter worked alone, television enjoys the partnership of an older and firmly established companion art. By means of sound broadcasting, television has a voice to speak the story which it is acting. Synchronized sound broadcasting for television is simply a partnership of both arts—sound and sight.

THE PICTURES ARE SMALL.—No one denies that the home television reproduction of today leaves much to be desired,

A FLYING SPOT OF MAGIC

but so did the early broadcast receivers with crystal detector and earphones. The present pictures usually measure not more than an inch and a half square. They may be magnified by lenses in which pictorial imperfections become more apparent. And the brilliancy is proportionally reduced. Viewed through a shadowbox or peep-hole by one or two persons at a time, the performance is reminiscent of the early days of the motion picture when a penny-in-the-slot and the turn of a crank brought animated scenes before the eyes.

Judging from present technical standards, such thoughts as televising field events and pageants are fantastic but by no means impossible of realization in the future. To one who has seen sound broadcasting develop from the faint whisper of the human voice to a full symphony orchestra, anything is possible under the destiny of modern research.

Professor Elihu Thomson predicts that the whole world may some day be able to see a total eclipse of the sun through the medium of television.

"Though direct observation of a total eclipse is necessarily confined to the dark tract of the moon's shadow, television may bring us from a distance images of the sun in eclipse," said Thomson lecturing in London. "This prediction may be fulfilled in August, 1932, when an eclipse cuts across New England; but technical development of television broadcasting may not then be sufficiently advanced."

IMPORTANT FORWARD STRIDES.—Zworykin in his laboratory at Camden, N. J., is privately showing his television receiver.

"Zworykin asked us if we wanted to view the images over a wire line in the laboratory or at an outpost five miles distant to which radio would carry the faces," said a New Yorker privileged to see a demonstration. "We chose the outpost and went by automobile to that point. There we had a pre-review. The picture detail was excellent. It was clear, in fact, most uncanny."

Television needs more than a good transmitter and a good receiver. They can be built and controlled by man but not so with the invisible waves. What waves are right for television? That is an important question. Alexanderson has turned his attention to this part of the problem. He is studying wave propagation. Germany reports that some of his images released from wires in the Mohawk Valley have been plucked from space near Berlin.

A Blurred Jumble of Galloping Horses—June 3, 1931

Cameronian prances out on the track at Epsom Downs a favorite. With this gallant horse, vying for the lead with Gallini, Orpen, Goyescas and others, a television camera looks down on the scene for the first time and thus the 1931 English Derby is the first to be televised.

It is called a "telecast." The parade of the horses before the start of the race and the crowds around the winning post are seen by distant observers. 'Tis true the pictures are not always clear. Static and other interference make the television scene at times appear as if viewed through a snowstorm. But, nevertheless, they know it is a horse race televised in the open air where artificial illumination is impossible.

Under the editorial caption, "Viewing the Derby by Television," the New York *Evening Post* comments as follows:

> While 750,000 were on hand at Epsom Downs to see Cameronian lead the field in the English Derby, a small and select group of spectators saw the finish of this historic race in their own studies. It is true that Cameronian's triumph was indistinguishable to these stay-at-home racing fans in a blurred jumble of galloping horses, but so must it have been to a tremendous majority of the thousands at Epsom Downs.
> Moreover, while the members of the stay-at-home group missed a great deal of the excitement which pervaded the course itself, they also avoided the crowd and tedious

A FLYING SPOT OF MAGIC

journeys to and from the race. It was, of course, through television that this privileged group of Englishmen watched the Derby in peace and quiet.

However faulty the transmission may have been, the experiment afforded a taste of what is to come. Slowly but steadily television is making its way. It is still in the stage which characterized the first awkward experiments with moving pictures and is subject to narrow limitations, but scientific workers engaged in its development differ in their forecasts only with regard to the time when it will be commercially practical. Television will some day be a commonplace and we shall view Derbies or football games or Presidential inaugurations as we now hear them over the radio.

On Derby Day, John Baird brings a strange wagon to Epsom Downs. It resembles a van from a gypsy caravan. But a mirror projecting on the back of the rear door of the vehicle gives the onlooker a clue that this might be the contrivance of a magician rather than that of a wandering gypsy.

A MIRROR WITH REVOLVING EYES.—The mirror dispenses with the necessity of revolving the "eye" on an axis in order to follow the race and to "see" various sections of the track. This television looking-glass is on a hinge so that it can be turned at different angles. Inside the van is a revolving drum the periphery of which is equipped with thirty small mirrors. It scans what the big mirror reflects. As the drum revolves the mirrors cause a strip of the scene to pass through a lens aimed at the photoelectric cells. They turn the light into electricity. There being thirty mirrors, the race track picture is cut into thirty adjacent strips. The process is repeated twelve and one-half times each second so that the distant observers, some of whom are fifteen miles away, see a complete picture.

Telephone wires connect the van with the television control unit at Long Acre from which point the signals are forwarded to Brookmans Park for broadcasting by the British Broadcasting Company's transmitter tuned to the 261-meter channel.

Television spectators report that they see the horses and jockeys parade before the start. They hear the clamor of the race track. They see men and women walk across the foreground little realizing that their actions are being watched by an eye that televises. They see the leaders dash past the finish post quite clearly but they cannot be identified individually.

This broadcast from Epsom Downs is heralded as a crowning event for television, one that foreshadows its possibilities. It demonstrates that outdoor events can be televised in sunlight without the glare of artificial lamps. This English Derby is just the start of a greater race—world-wide television of news events and scenes of action.

Color Music From an Organ—June 4, 1931

Problems of the television showman are simplified in the beginning by the fact that the performance is in black and white. When color is added to the ethereal pictures care will have to be taken that the tints are synchronized with the music. The eye and ear must not clash. This will be an important factor so far as entertainment value is concerned. The engineers assert that one of the main tasks is now in the creation of projection apparatus to permit the rendering of color in a form as appealing to the eye as a symphony is to the ear. The artistically inclined lighting expert foresees a new opportunity in television.

Already an automatic color organ, called a by-product of radio, has been developed to produce colors by means of music and to synchronize colors with music. Television in years to come may give it a wide field of usefulness.

"It seems that to correlate sound and color is at once impossible of solution," said E. P. Patterson, the engineer who

A FLYING SPOT OF MAGIC

discussed the color organ at a meeting of the Institute of Radio Engineers. "In spite of this, pleasing results can be obtained because æsthetic enjoyment is not based on formula. With color the eye perceives three factors—hue, degree of saturation and brightness."

HUMAN REACTIONS TO COLOR.—It is generally recognized that colors exert a profound influence over the majority of people. The following table by Luckiesh gives a series of colors with the commonly associated reactions. It may be useful to the television showmen when they can utilize tints.

Red—warm, exciting, passionate.
Orange—warm, exciting, suffocating, flowing, lively.
Yellow—warm, exciting, joyous, gay, merry.
Yellow-green—cheerful.
Green—neutral, tranquil, peaceful, soothing.
Blue-green—sober, sedate.
Blue—cold, grave, tranquil, serene.
Violet—solemn, melancholy, neutral, depressing.
Purple—neutral, solemn, stately, pompous, impressive.

"A method of harmonizing color and music is to assume that the bass notes of the drum indicate an effort on the part of the composer to create a stirring effect and hence a red color," said Patterson. "In practice red may usually be assigned this position. The other colors, however, represent a more complicated problem. It is possible, with special arrangements, to obtain a most sensitive control, the colors following practically every change in the music. However, violent fluctuations tend to become objectionable. Where extremely rapid changes are required, incandescent lamp filaments should not be too heavy on account of the time delay in heating and cooling. While we have roughly determined the color of the lights to be employed, the success of the presentation depends greatly upon the manner of light projection and also on the introduction of some moving patterns, which serve to relieve the possibility of monotony.

TRICKS OF THE TRADE.—"There are a number of effect machines to produce clouds, waterfalls, rain, etc. These, for the most part, consist of a revolving or painted disk in front of a spotlight. These spots may be directed on a curtain, and used in conjunction with ordinary border and footlights as found in the theater. Elaborate lighting schemes are coming into prominence where bare walls are painted by color patterns and projected pictures. These systems serve to focus public attention on the lighting art, and lend themselves to the easy adaptation of color music. In the creation of patterns, moving or still, care must be exercised in avoiding too definite a structure. The imagination is important in giving æsthetic enjoyment which cannot be realized to the fullest extent when the pattern is too concrete in form, even though it may be beautiful in design."

FACES THAT INSPIRE PREDICTION—JUNE 9, 1931

There is a roar like a huge printing press getting under way, but not quite so loud. It is television coming to life. Sanabria is manipulating the switches and gadgets that complete the copper pathway over which electricity rushes into his television machine at one end of a ballroom in a Chicago hotel. At the other end of the room is a six-foot screen upon which all eyes are focused.

A spot of light, about the size of an orange, flashes on the darkened white sheet. Slowly it begins to move to the right and off the side of the screen. Other spots whiz across, one above the other. A motor that controls their destiny is gaining speed and the cluster of light spots moves faster and faster. Now they appear as illuminated lines instead of round daubs of light. The screen is streaked and resembles one of those large transparent washboards with the parallels of light from top to bottom recalling the board's corrugated surface. Suddenly a face stares at the audience. It is blurred. Nevertheless, it affords a distinct glimpse of what is coming. Sanabria turns a knob or two and the focus is improved. The face is clear. The crowd applauds. The large-sized head on the screen makes a bow.

An observer is inspired to predict that shopping by television is likely to play a part in the scheme of living in the not too distant future.

"Just as the broadcasting stations send out shopping, diet, health and other talks of interest to women, television will transmit the images of the wares from stores and shops," he said. "Women will tune in by television before they begin their shopping tours. Television shopping will save time and energy, allowing more opportunity for other pursuits. Women will have more time to relax. They can see the bargain counters from home. They can see the wares by radio and place the order by telephone."

FANTASTIC THOUGHTS.—The question is where are all the shops going to get wave lengths. And what a mix-up there would be if hundreds of grocery stores, butchers, hardware stores, bakeries, drug and candy shops adopted the same idea!

Another visionary person believes that industrial corporations will hold their board of directors meetings by television. The chairman will call to order a meeting of electrical personalities, and they will discuss the affairs of the corporation as if all were present in a room. He even expects that documents will be televised and comments broadcast.

But what about the millions who might eavesdrop? Television must be more secretive than radio broadcasting before this dream can come true.

This same person anticipates that fifty years hence the Premiers of France and Italy will talk and see across the water and save weeks of time, now necessary when they make a trip to see the President of the United States. He contends that our grandchildren will wonder at the quaint custom of the past that made it obligatory for members of Congress to convene in Washington. He foresees the time when they will debate and pass laws by television. Filibustering will have no terrors. Tuning out will be a simple matter.

It will be many a day before statesmen trust the all-pervading radio and television to handle their diplomatic communications. Space cannot be trusted with secrets and important plans. Open discussion is not always desired. And the day when Congress will meet by television, well, that calls for a vivid imagination.

A Prophecy That Fell Short—July 18, 1931

Television is forging ahead in England despite the fact that radio as encountered abroad by H. G. Wells, several years ago, was found wanting. The English novelist observed that the invisible audience was disillusioned and bored to death while lifeless aerial wires sagged between chimneys as useless as barbed wire entanglements abandoned after war. He wondered if any indefatigable listeners stuck to the ethereal amusement for more than two weeks and if so he thought that they must be "very sedentary persons living in badly lighted houses or otherwise unable to read" or they have "no opportunity for thought or conversation."

That was in 1927. The English listeners look back to June 15, 1920, when Melba broadcast from Chelmsford as the pioneer performance that revealed the possibilities of entertaining by way of the microphone. Then two years passed before a regular broadcaster, station 2LO at London, went on the air, November 14, 1922. All owners of radio receivers in the British Isles are now forced to acquire a license from John Bull so he knows from day to day whether or not his radio family is growing or dying. Today there are more than 3,500,000 licenses issued, despite the dire forebodings five years ago.

A FLYING SPOT OF MAGIC

THE KING PROVES A POINT.—The sagging aerial wires apparently have been tightened. Radio broadcasts have shown in many ways that antenna wires are a bit more potent than abandoned barbed wire on No Man's Land after the war.

What about radio today? The radio that triumphed at the Naval Arms Conference in London, on January 21, 1930, when King George V spoke into a golden microphone to the greatest and most cosmopolitan audience that ever listened simultaneously to the voice of a monarch. The broadcasters estimate that on that occasion 100,000,000 tuned in!

President Hoover and his "medicine ball cabinet" gathered round a loudspeaker at the White House. France reported that the radio speeches calmed the press by their sincerity and goodwill. Australians recognized His Majesty's voice. Manila picked up the 25-meter waves from London. Japan was a trifle disappointed. Reception was "loud, roaring and brassy," because a pianist at a Russian station unceremoniously crashed into the Japanese ether. Listeners at Jungfraujoch, 13,000 feet above sea level in the Alps, were in tune.

Germany enjoyed excellent reception of "a veritable babel of voices that swamped several millions of radio fans for two hours with a technical perfection that left little to be desired." Stockholm, Paris, Vienna, Basel, Budapest and Rome picked up each syllable distinctly. All parts of India eavesdropped on the delegates in the House of Lords. Radio did not miss a single nation as it sprayed the surface of the earth with the messages of peace and goodwill. If radio can do that with a voice, is it not possible that some day it will do likewise with a face?

THE FATE OF THE PIONEERS—AUGUST 26, 1931

When one looks back in the files that have preserved the radio programs of 1921 and 1922, he finds that in most cases the names of the pioneer entertainers are strange. Few of them carried on with the development of the art. Many of them went on the air just for the novelty.

For example, those who donned the earphones on May 25, 1922, and adjusted the crystal detectors, may have heard WJZ begin its Sabbath broadcasting at 3 o'clock in the afternoon with chapel service from the Episcopal Church of Paterson, N. J. This was followed by a musicale featuring Louise B. Wilder, Lucille Bethel and Mabelanna Corby. Then came the literary vespers by Edgar White Burrill. Readings and phonograph recordings by Ralph Mayhew were on the air at 6:30 o'clock, followed by Sandman stories by Kaspar Seidel. "Business on the Upward Trend" was the topic discussed at 7:20 o'clock by J. H. Tregoe. P. W. Wilson faced the microphone to report the latest foreign news. Alfred Sgueo, violinist, gave a one-hour recital. Music by the Orpheus Quartet of Newark furnished the finale of the day and WJZ signed off at 10 P.M.

The names of the pioneers have been supplanted by trade names and names of advertisers who sponsor the entertainment. So it may be with television when one looks back a few years hence to the following performances broadcast by W2XAB, New York on the 107-meter wave while W2XE handled the sound on the 49.02-meter channel:

August 24, 1931

2:00—6:00 P.M. Experimental sight programs. Demonstration of card station announcements and drawings of radio celebrities.

8:00 P.M. At Home Party, an informal studio gathering showing set-up for party and large group of people. Research in televising whole scenes, utilizing groups as a background.

8:30 P.M. Dancing in the Dark, featuring Natalie Towers in a series of television waltzes. Five different lens pick-ups. Test with silver background curtain.

A FLYING SPOT OF MAGIC

8:45 P.M. Television Crooner, Doris Sharp, dressed entirely in red. Experiment to show effect of color in television pick-up.

9:00 P.M. How the Best-Dressed Girl in Radio Should Look—Mary McCord wearing the latest fashions from Paris.

9:15 P.M. Tap dancing demonstration featuring Jack Fisher in a song recital with self-accompaniment on the violin.

9:30 P.M. Recital by Charlotte Harriman, contralto. Use of sun-tan and white make-up.

9:45 P.M. The Bon Bons, quartet in costume.

10:00 P.M. Twin violin demonstration featuring Virginia and Mary Drane with Carol Seaman. Long shot pick-up. Half-length focus.

10:15 P.M. Helen Nugent, contralto.

10:30 P.M. Dramatic Readings with modernistic background featuring Alice Raff.

10:45 P.M. The Singing Vagabond, Artells Dickson. Character songs and stories.

August 25, 1931

2:00—6:00 P.M. Experimental sight programs. Demonstration of card station announcements and drawings of radio celebrities.

8:00 P.M. Ernest Naftzger presents following artists—
Girls' Trio—Dorothy, Alice, and Jean
Islay Benson, English Character Artist
Louis Bia Monte—Saxophonist
Ethel Parks Richardson, Hill-Billy songs.
Test to determine clarity of single artist pick-up as contrasted with three or more.

8:30 P.M. Teddy Bergman, Television's Clown.

8:45 P.M. Pantomime Demonstration featuring Grace Voss in three pantomimes. Long shot pick-up with white screen background.

9:00 P.M. Puppet Follies presented by Peter Williams.

9:15 P.M. Television Taps—Tap dancing specialties by two five-year-old boys; long shot pick-up attempting to get in whole figures.

9:30 P.M. Exhibition Boxing Bout. Three demonstrating possibilities of broadcasting boxing matches by sight and sound.

9:45 P.M. Chess Playing Demonstration featuring Edward Lasker, Major Ivan Firth and Gladys Shaw Erskine. Test shows chessboard and how a certain championship game was played.

10:15 P.M. John Brewster, juvenile actor in novelties.

10:30 P.M. "Waltzing Through the Air." Natalie Towers dancing by television. Close-up and long shot.

10:45 P.M. "Songs of Spain," featuring Soledad Espinal.

And so the show goes on! What will they think of these broadcasts when 1950 arrives? They will probably smile to think of Natalie Towers waltzing to the tune of "Dancing in the Dark," while the technical experts were just as much in the dark regarding some of the problems that look so simple from the 1950 point of view. By that time the tap dancers ought to be in demand, and television may be able to do justice to portrayal of the best-dressed girl in radio. Studio boxing bouts will have passed from the air as the regular championship fights are picked up at the ringsides. And the Puppet Follies will give way to the glorified girl, while entertainers galore seem to waltz through the air.

TELEVISION ON A BEAM OF LIGHT—DECEMBER 22, 1931

Lured by ultra-short waves, Alexanderson has decided to experiment with television traveling on a beam of light. He successfully demonstrates in the laboratory that the images will follow a light ray a billionth of a meter in length, thus opening the way to a new field of research in which he sees numerous possibilities.

Instead of feeding the electrical impulses into a radio transmitter, they are modulated into extremely high frequencies on a light beam from a high intensity arc. The beam is projected the length of the laboratory where it strikes a single photoelectric cell which transposes the modulated light waves back into electrical waves. The electrical impulses reproduce the image by means of an ordinary television receiver.

"The work thus far is highly experimental," said Alexanderson, "but some day we may see television broadcast from a powerful arc light mounted atop a tower. The modulated light waves will be picked up in homes by individual photoelectric cells, instead of by an antenna. Light broadcasting may have the same relation to radio broadcasting as the local newspaper has to the national newspapers. These light waves can be received at relatively short distances, possibly ten miles. Each community could have its light broadcasting system. The logical progress of this development is in the exploration of still shorter waves than are found in the radio spectrum. That takes us into light waves which we know travel in straight lines. Furthermore, they can be accurately controlled by such optical means as mirrors and lenses.

"When it was decided to take up experimentation on this subject Dr. Irving Langmuir of the research laboratory was consulted about the probabilities of being able to modulate a source of light at the required high frequencies of from 100,000 to a million cycles. Dr. Langmuir, who has done much research work with arcs, believed that this could be accomplished by using a high intensity arc. It was concluded that a most desirable light would be a high intensity arc of the type where the light comes from the arc rather than from the crater. In the 10-ampere arc lamp used for the first test most of the light comes from the crater, and comparatively little light is in the arc. The lamp was used in such a way that the light from the crater was eliminated, and the arc used was, therefore, quite a weak source of light. The current from our standard television pick-up was superimposed upon this arc, and the light from the arc intercepted by a photoelectric tube at a distance of 130 feet. The photoelectric tube was then used to control the regular television projector. The image transmitted in this way had the same sharpness of detail as the one ordinarily obtained without the interposition of the light beam."

Jenkins Tries Lantern Slide Scanning—March 1, 1932

The television screen up to now has been swept by daubs of luminous "paint", which picture the image. Small screens are generally used because of the gigantic task a single light spot is called upon to perform in order to illuminate a large screen of theatre size. That is why some engineers contend that as long as only a tiny light-brush is available to sweep across the screen, television images will be confined to a small area.

C. Francis Jenkins has been hunting for a new principle. In the Yale Scientific Magazine he reports that he is substituting "persistence of picture element for persistence of vision." His new method does not involve a rapid transversing of the picture area by a single spot of light in adjacent parallel lines. The entire picture is on the screen all the time instead of only a single gyrating dot of light.

Broadly, the new method consists in utilizing the incoming radio signals to build up a picture in the path of a light beam projected on a screen. There is a fixed lantern slide upon which the objects move instead of being stationary as on a magic lantern slide. The picture on this animated slide is scanned or formed thereon by electrical rather than photographic means. The slide replaces the flying light spot employed in other systems.

The mechanics of the method consist of dividing the picture area of the lantern slide into sixty imaginary lines of sixty dots to each line and changing the chemicals in the gelatin coating of the plate to attain the fading of an image and its replacement by a like image every fifteenth of a

A FLYING SPOT OF MAGIC

second. The prepared slide is put into a projecting lantern equipped with a light source.

In the receiver, in front of, closely adjacent and parallel to this animated lantern slide, a suitable transparent scanning disk is mounted. It has sixty wire terminals on its face to distribute the incoming radio impulses along each of the sixty lines on the slide. Jenkins estimates that using these wire-like nerves about 3,600 times as much light can be utilized.

Thus Jenkins seems to have accomplished the impossible. He has arranged a transparent lantern slide plate with a sensitive surface so that it will become transparent or opaque in response to rapid changes in light. Areas of crystal clearness and darkness move about in an ever changing pattern and great rapidity, reproducing the picture flashed by the transmitter.

"When the transparent scanning disk is brought into synchronism with the analyzer at the transmitter, the incoming radio signals form spots on the lantern slide", said the inventor. "Each spot is an element of the picture of the person or scene being televised at the transmitting station. All the spots are put in their proper places in a tiny fraction of a second.

"But as rapidly as each spot is put on the plate by the incoming radio signal it begins to fade. The fading time is one-tenth of a second and a complete respotting occurs every fifteenth of a second. Obviously, each spot is in its place all the time in the stationary part of the picture. If, however, a particular group of spots form a moving part of the picture, for example, a speaker's arm in gesture, new spots will be formed in successively new locations as the arm moves to new positions, and the old spots fade quickly.

"The projected picture on the screen is, therefore, exactly like the usual lantern slide picture except that it has motion; or like a motion picture except that it is made up of changing elements instead of changing picture frames of a film. Incidentally, the elementary picture dots are so blended that they are as inconspicuous on the theater screen as are the picture dots of a newspaper illustration."

This method is described as being somewhat analogous to the three-element vacuum tube in which a little current on the grid controls the flow of a relatively large amount of current. The feeble radio current in this television method is not the light source, as it is in other systems, but the radio impulses are used to block out, in simultaneously-acting elementary areas, a beam from a powerful light source. Therefore, no interrupting shutter is utilized. Twice as much light reaches the screen as in a motion-picture projector where a rotating shutter cuts off half the light.

Jenkins contends that with this system any size screen can be adequately lighted for large gatherings, to accompany a synchronous voice-amplifier. He says that a small incandescent lamp is quite ample for home radiovisors, synchronized with the loud-speakers now in use.

PART V

A GLIMPSE AHEAD

TELEVISION'S FUTURE
Editorial
The New York *Times*—September 28, 1931

Such has been the progress made in television that at the opening of the recent radio show Mr. David Sarnoff predicted that next year would witness the establishment of what he termed "the theater of the home." What we shall see on the screen of the partially darkened living room will be a living image about six by eight inches in size and about as well defined as a newspaper half-tone picture. Synchronized with the broadcast voice, the electrical counterfeit presentment will sing, talk and smile.

Surely these small images will but whet the appetite. A performance of "Parsifal" at Baireuth visible and audible in New York, with singers as large as life—why not? Television lends itself to such imaginings. No engineer will deny that ultimately they will be realities. But give him time. Consider what television means even on a small scale, he reminds us.

In a motion-picture theater we see a dozen whole pictures in a second, and because our eyes cannot separate one from another we obtain the illusion of continuous motion. But a televised image consists of points of light alone. Several hundred thousand of these must be assembled every second to fool the eye into accepting them as a whole. The more points the better the picture. A million a second would give us the detail of a good photograph.

It is hard enough to obtain these many brilliant points on a small screen to reproduce only the head and shoulders. To see the full figure on the distant stage, there must be more intense light than the engineer can now generate, and a more flexible distribution of it. There must be something far more sensitive than the photoelectric cell or "eye" of today which converts the points of light at the transmitter into electric impulses and reconverts these at the receiver into an image. There must be an optical system more effective than anything thus far devised to collect and concentrate light rays. Ordinary broadcasting is child's play compared with television on such a scale.

State the problem and its solution seems impossible. Yet the history of invention is full of "impossibilities." The television of today was just such an impossibility only ten years ago. With a half dozen research organizations here and abroad devoting their energies and technical resources to electrical communication, who will deny that we shall see across space as effectively as hear across it, that we shall be electrically present at great public festivities of the future, that the chief dramatic and operatic performances of New York, London and Paris will be retailed, as it were, to 50,000 theaters scattered throughout the world for the benefit not of a few fortunate travelers but of whole nations?

Chapter Fourteen

TELEVISION'S COMMERCIAL DESTINY

Television scintillated on the mind of man long before it flashed on a screen. The human race has long anticipated that some day science would make it possible that man "looketh to the ends of the earth, and seeth under the whole heaven." What medium except radio could fulfill such hopes? Television is the crystallizing of the dream.

Charles H. Sewall, writing on "The Future of Long Distance Communication" in *Harper's Weekly*, December 29, 1900, revealed that more than thirty years ago there were some who foresaw television:

> The child born today in New York City, when in middle age he is visiting China, may see reproduced upon a screen, with all its movement and color, light and shade, a procession at that moment passing along his own Broadway. A telephone line will bring to his ear music and the tramp of marching men. While the American pageant passes in the full glare of the morning sun, its transmitted ray will scintillate upon the screen amid the darkness of an Asiatic night. Sight and sound will have unlimited reach through terrestrial space.

All Are Not Hopeful.—A quarter of a century has seen research move closer to the possibility of that prediction coming true. The Asiatic has yet to be thrilled by the sight of the Great White Way coming to him through space. Yet, the child born in 1900 may see the dream a reality. Despite the fact that television has moved more than thirty years nearer the goal, its destiny in 1932 is subject to conjecture and wide diversity of opinion. For some it sparkles optimistically. Others view it with pessimism. They do not expect to see much or far by radio vision.

Out of the depression that fell upon the world in the autumn of 1929 the cry of television is heard as never before. The youthful radio industry inexperienced in business cycles, the curves of which turn downward, is hard hit by adversity. Television is heralded as the savior. Some call it mere ballyhoo. They look at television as merely a rose that will fade in the lapel over the aching heart of the radio industry, suffering the pangs of its first great business illness.

RADICAL DISCOVERY NEEDED.—It was in July, 1926, that Dr. Lee de Forest was asked what he foresaw for the future of television.

The inventor shook his head as he remarked, "I am very skeptical as to the future of television—not from a theoretical standpoint but from a commercial. I think that with our present knowledge of physics and natural phenomena, an operative system of television can exist only at an expenditure of an enormous amount of money and after long research. The equipment involved would be exceedingly expensive, delicate and require most expert manipulation. It can, therefore, obviously not become a popular instrument to be placed promiscuously in thousands of homes.

"It might be possible, granted there was the necessary expenditure of money, to project a prize fight from New York to Chicago or San Francisco so that it could be seen on the screen in large auditoriums in distant cities, but the equipment making this possible would cost so much to build and maintain that I do not believe the large corporations, which alone have resources adequate for this problem, will feel justified in making the necessary investment over the term of years required.

"At the same time it is conceivable," said de Forest, "that some one at any moment may come across a radically novel discovery in physics which will make this problem simple. Such is entirely in the speculative realm, however. I am not particularly interested in going into speculative trances, based on what we know nothing of, and prompted only by hope and hootch."

TWO DIFFERENT VIEWS.—A radio man steeped in television research in Chicago sees no great future for the art as far as home entertainment is concerned. He says that he cannot stretch his imagination to the day when 1,000,000 television sets will be in American homes. Nevertheless, statisticians of a broadcasting organization estimate that by 1942 there will be 11,000,000 television sets in the United States. The Chicago engineer is also at a loss to foresee what sort of a sound-sight program could be broadcast to please a fastidious public. He is told that he should not worry because that end of television belongs to the showman and not to the technical engineer.

Three months later a noted radio engineer, who won much fame from inventions of facsimile broadcast apparatus, is interviewed in New York. He has sent facsimile pictures and messages far across land and sea.

"Do you think there is any future for television?" he is asked.

An emphatic "No" is his answer without the slightest deliberation. He is sure of it.

"How can it have a future outside the laboratory?" he said. "Suppose you were called upon to develop a motion picture for the theater, with no radio or wire transmission involved, and all you had to work with was a single spot of light. And that is the case with television. The pictures at their best are streaked and freckled, blotched and mangled."

This pessimistic engineer is reminded that his facsimile pictures were difficult to recognize in the beginning. Yet, today he has given up facsimile research because he feels that the apparatus has reached a point so near perfection that all improvements from now on must be made in radio circuits. The pioneer broadcasts of music were crude and distorted with the melody destroyed by static bombardments. Nevertheless, the public enjoyed them. They bought thousands upon thousands of receiving sets. The tone quality and the types of performances improved rapidly. So might television when it is given a chance to stretch its wings outside the laboratory under actual operating conditions. Then the engineers nurturing it would learn much more about it and possibly make greater strides, as they did with sound broadcasting.

It is true television is more complex. Its life may revolve around a single, flying spot of light, but there is no reason why greater things cannot be developed from that nucleus. The camera with its eagle eye and lenses developed from a pin hole. The electron is a tiny speck of electricity but wonders are performed when millions of them get together. It might be that way with light spots. The fact that television begins with a spot of light does not mean that it will stop there. Modern scientific research will eventually prove that it can take the apparently insignificant spot of illumination and with it see to the ends of the earth.

"Well, you are an optimist," smiles the engineer. "Keep to your faith and some day you may be right."

A POWERFUL FORCE LOOMS.—Television, because of its far-reaching aspects and its magic possibilities that can influence many affairs in this world, is likely to bring with it a new era in international relations. It is bound to have a marked effect on home-life, on education, business enterprises, religion, literature and to play a diversity of rôles in art, science and entertainment. It will cast its spell over the theaters. It will influence the newspapers, magazines, and many other agencies that play a part in everyday life. The advent of the television era can be compared in importance with the arrival of the electric light that dimmed the glory of candle and kerosene lamp; with the arrival of the automobile that relieved the horse, sped up travel and introduced good roads that linked the farm with the city.

TELEVISION'S COMMERCIAL DESTINY

TIME WORKS WONDERS.—When wireless began under the tutelage of Marconi it was difficult for many to believe that dots and dashes could be sent across the sea without the use of wires. But there were some whose imagination carried them afar to predict the day would come when wireless would carry voices, music, pictures and possibly motion pictures that talked. Perhaps they foresaw all this for the year 2,000. In 1900, it took a long stretch of the imagination to look ahead to the day when a young man would fly across the Atlantic from New York to Paris in thirty-three hours, and that four years later when he flew to Japan his voice would be heard throughout America as he was fêted at a dinner in Tokyo. Yet that happened to Lindbergh in 1931.

"With the advance of radio and aircraft," said Lindbergh, "the mystery of distance no longer exists. . . . We have come to Japan over the top of the world, and when we were near the North Pole we discovered that the people in our country were not rightside up and those in yours upside down but that both were really walking at the same angle. We discovered no line separating a green country from a purple one on our maps. The ideas which we have inherited from past ages become insignificant when we look at people from the sky and speak to distant people through the radio. I do not know what effect aircraft will eventually have on the world, but I have great confidence in its future. You must not, however, expect too much in one generation."

Scientific progress does not come in a flash overnight. Twenty-five years passed and Marconi celebrated his silver jubilee in wireless before KDKA went on the air as the pioneer broadcaster.

Boys who first saw the light of day when Marconi was sending early signals in Italy played a vital rôle in the later developments that girdled the earth with voices and melody. Boys in the cradles, in the kindergarten and making mud pies in their backyards today will no doubt get on the track of television as the years roll on and they will make the dreams of today come true. Television is just beginning. It is a gigantic task. It cannot all be worked out in one generation. The Edison of television may be unborn in 1932. He may be just learning to creep in a rural home. One thing is certain: television offers opportunity—it is a promised land for youth endowed with a scientific mind or talent in research and showmanship.

THE TELEPHONE MAY "SEE."—Television may educate future generations to expect to see people who telephone to them. They may look upon the present telephone system as something that should be improved by the addition of sight. When the silent film took on a voice and became a talkie there were many skeptics and many who preferred the silent drama on the screen. But it was not long before the talkie revolutionized the motion picture. So television may influence the telephone. The next generation may want to see the speaker at the other end of the line because seeing by radio creates the desire. That may be the reason why the Bell Telephone Laboratories are keeping abreast of television and are experimenting with two-way television from one telephone booth to another. Today the sightless telephone is all that is desired. But those who follow in an era of radio vision may not be content to converse without seeing.

TELEVISING THE CLOCK.—Today when a telephone subscriber desires he can call a certain number and hear the correct time struck. He can turn on the radio and hear an announcer reveal the exact location of the hands on the face of the clock. In years to come it may be different. There may be a television clock. It may be the master clock of the nation in the Naval Observatory in Washington. Throughout the day and night, on constant duty will be a television eye focused on the face of that timepiece. It will always be on the same wave length. That will be the time wave.

When the owner of a television receiver wants the correct

time he will merely turn the dial to that wave length and the face of the clock will be right there visually to announce its own story.

England may have a television time camera trained on the face of Big Ben atop the House of Parliament, and so the famous timepieces will be given a new long distance range. Millions will see their hands brush away the minutes, instead of a few who pass in the street.

ADVERTISING BY TELEVISION.—There seems to be no end to what television may do. Mariners in mid-ocean will watch prize fights on shore as the ringside scene travels to them from New York or Chicago. The roped arena will probably be one of the first successful sports events on the television screen because it is not spread out like a baseball diamond or football gridiron. The ring is twenty-four feet square and there are only two contestants for the radio camera to keep its lens trained on.

Advertisers will demonstrate their products, in fact, they will help finance the television performances in much the same way as they do broadcasting. Advertising characters that have long been stationary on cereal boxes, coffee cans and wrappers will have life instilled into them by television, because some of them will be enrolled as performers.

An insight to what television will be like when the commercial sponsors grasp it as an advertising medium is found in this announcement made in connection with a 1931 program:

> The first million-dollar television broadcast will be staged at W2XAB, New York, on Tuesday night September 8, at 10 o'clock, when rare and historical gems from Cartier's vaults worth more than that amount will be on display before the photoelectric eyes.
> Natalie Towers, original television girl, will wear the gems. Ranging from pearl necklaces to emerald rings the whole gamut of jewels and precious stones will be covered.

Special emphasis will be placed on engagement rings, their evolution and fashions today.

The display will start with a short pictorial history of the engagement ring. The history of stones used to plight troths will be told in words, while Miss Towers displays the romantic circles. The program will include a showing of other jewels—pearls, diamonds, rubies, emeralds—historical and modern, and many pieces of rare art from the private collection.

If that can be done with precious gems, think what an announcer can accomplish in a coast-to-coast television demonstration of a new automobile as he points out the salient features while the shining chassis revolves on a turntable in front of the television eye.

Think of the possibilities and great response an advertiser might have should he conduct a "guess who" contest by television. Only parts of faces of prominent actors will be shown and the invisible audience will be asked to identify them, and those who guess correctly will win the prizes.

Beautiful girls will be in demand for the Follies of the Air. They will be called upon to play leading rôles in the television broadcasts that advertise everything from coffee, that their winning smile reveals is "good to the last drop," to the latest hats, shoes, dresses, pajamas, bathing suits, cigarettes, candy and soup. There may be an Arrow Collar Quartet and a General Electric tap dancer, while General Motors contributes the world's greatest troupe of acrobats and Ford sends in a famous ballet corps. It will be easy for the announcer or a pretty girl to point out the date on the can of fresh coffee.

Television will save many a descriptive word on the air because the pictures will tell the advertising message quickly and effectively. Television will revolutionize the system of sound broadcasting that has taken more than ten years to evolve. Actors, singers, musicians, dancers, acrobats, magicians and entertainers of all sorts will flock to the glow of the photoelectric cells as thick as insects around an arc light on a country street. Television will give the arts a new medium of expression; talent a new opportunity.

Airplanes will carry television monocles which will enable the pilots to see through fog and darkness so that they may land safely. And ships at sea like the serpents in the story books will see far across the waves, far across the horizon and the curvature of the globe.

A NEW MEDIUM OF FRIENDSHIP.—Images of statesmen and their friendly gestures will mingle among the nations. Television will usher in a new era of friendly intercourse between the nations of the earth. Current conceptions of foreign countries will be changed. Television will perform in this respect in much the way that Lindbergh saw aviation creating new friendships when he said to the Japanese in Tokyo:

"We have come to Japan for an opportunity of meeting your people and learning a little more of the country which in our schooldays was known to us in America as being on the other side of the world. When we were children, we thought of Japan as a land filled with people who were different from us as though they lived on another planet. We marveled at their ability to walk upside down and that they kept from falling off the earth altogether."

And so television will enable the inhabitants of the earth, who do not have the opportunities of travel, to see how their fellow men live on the other side of the globe. They will learn to enjoy their music, drama and national scenes. Suspicions will be obliterated. New friendships will result. No one will see the other nation always "walking upside down."

When the Japanese Premier, the late Hamaguchi, broadcast the first message of goodwill to listeners in the United States his voice was remarkably clear despite its long flight by short wave across the broad Pacific to the California shore He opened a new era in international relations between the United States and the East. Then the airplane of Lindbergh flew over for a visit. The next link in the chain of friendship may be television—when Japan will see America and Americans will see the Japanese.

Is the Theater in Danger?—What effect will television have on the theater? Will the public go to the movies if they can see the films and news reels flash on screens amid the comfortable atmosphere of their home? Will they go to the theater to view a stage performance or to an auditorium to see and hear the opera?

The stage, screen and opera will endure. The leaders in radio are of this opinion else they would not plan great theaters for the actors and opera stars in Radio City. The theater, the cinema and the opera will probably be more spectacular in this new setting. They will be part of television. All will go hand in hand. No matter how elaborate the radio show, it will not keep people away from the theater, where the entertainers are seen in person untouched, unblemished by the elements that may attack them on their flight through space. Television could probably begin successfully in the home as a more or less peep-show, but in the theater, if it is to play a part there, it must compete with the motion picture.

Hollywood of the Air.—When the theater complained of a dearth of material; when playwrights said they had used up all the stories and plots of generations past, along came the motion picture to work them all over again. And when the silent films exhausted the dramatic themes, the talkies arrived at the opportune moment. And the old stories were still good. Now Hollywood says it is running out of ideas for screen adaptation. This looks like a sign that another step in the evolution of entertainment is not so far away. Television may be next to give the ancient themes a new avenue of escape and the actors further opportunities.

The old material will be freshened, old ideas and love

stories as old as the hills will be dressed up to amuse many millions over and over again. Everything that has been adapted successfully to the celluloid reel will find television a medium for greater triumphs. Television and the movies are destined to be related. There is a wedding of these two arts in the offing. Camera men, continuity writers, directors, actors, dialogue specialists, studio fashion stylists, decorators, swarms of artisans, carpenters, authors, painters, orators, scenic experts, musicians, electricians and mobs of extras comprise a small portion of the army that will hear the call of television inviting them to the Hollywood of the Air to participate in a great revival of all that has gone before on the silver screen.

The blood hounds that chased Eliza across the ice blocks in many a small-town opera house, and then across the movie screen, featuring *Uncle Tom's Cabin*, will do that same thing over again for the television audience. Ben Hur, who lived on the stage, raced in his chariot across the silent screen only to have his name blaze in brighter lights on Broadway when the sound picture brought him back, will race again in a television thriller, in a spectacle of nation-wide scope. The covered wagon will lumber through space as it did across the plains in the pioneer days. *The Birth of a Nation* and all the great pictures of the past will be seen again by television.

NEW FIELD FOR FILMS.—Television will be a good thing for the film business. Already the projectors or teleopticons, as they might be called, are handling reels of pictures instead of attempting to televise directly in studios or outdoors. It is more likely that dramas, comedies and geographic scenes will be photographed and prepared in advance of the broadcast on reels, which will be distributed to the stations in much the same way as electrical transcriptions or records are furnished the broadcasters. Television from film will be more perfect than direct photography, because mistakes can be corrected before the reels are completed. Much of the television show will be prepared in studios and presented as are the talking pictures.

Theaters may find it necessary to subscribe for a wire service that will bring them television news scenes, so that the audience can follow the events as they take place. There may be special television auditoriums that will feature the world series baseball games, the best stage productions, outstanding football games, championship bouts, international and national news events, boat races and hockey games.

However, there is much to be done in scientific research before a television picture is equal in size and clarity to the motion picture. Furthermore, it is one thing to televise a filmed studio performance adapted to broadcasting, and a much different proposition to televise an outdoor event alive with activity such as a football game. It is a rather difficult job to follow the gridiron contest on a movie screen and much more so to follow the plays as might be reproduced by any of the current television systems.

It will be many a year before the Harvard-Yale game is played with no spectators in attendance because they are all at home looking in. As long as the game is played there will always be the football crowd that prefers to be on the scene rather than at the screen. And so it will be with other sports events.

LOOKING IN ON SPORTS.—Some one remarked that to hear a prize fight by radio is like staying at home to look at a best girl's picture or to watch her in a home motion picture instead of being with her. The picture or film is a mere substitute. The enthusiast for football will not surrender the trip to New Haven or Cambridge for a motion picture by television. The world series fan will continue to go through the turnstiles. The theatergoer will continue to pay homage to the box office. Television will supplement. It will not supplant.

Everyone cannot crowd into a football arena, into a Radio

City theater, into the Yankee Stadium or into the roped ringside in Madison Square Garden, so what will it matter if the same scene travels across the countryside by television? It merely gives added millions an opportunity to enjoy the performance, but the man who pays admission sees it firsthand.

Broadcasting did not destroy the stage and motion picture as some predicted it would. They saw radio as a great monster threatening to claw the theater and the screen. Now some see television as the monster grown up and more ferocious. But it is away down the road of the future so far as the theater is concerned, and when it approaches it may not be as dangerous as it looks, because of its cooperative and supplementary features which will open new opportunities for all concerned.

WHAT SARNOFF FORESEES.—"The motion picture industry need experience no alarm over the impending advent of television," said David Sarnoff. "Transmission of sight by radio will benefit not only the radio industry; it will also provide a welcome stimulant, a pleasant tonic to all the entertainment arts. There will be no conflict between television in the home and motion pictures in the theater. Each is a separate and distinct service. Television in the home will not displace the motion picture in the theater.

"Man is a gregarious creature. Granting that we can develop 26,000,000 potential theaters in the homes of America, public theaters will continue to operate because people will go there in response to the instinct for group emotions and to see artists in the flesh. These are human demands which television in the home cannot satisfy.

"Television, when it arrives as a factor in the field of entertainment, will give new wings to the talents of creative and interpretative genius. It will furnish a new and greater outlet for artistic expression. All this will stimulate and further advance the art of motion picture production. The potential audience of television in its ultimate development may reasonably be expected to be limited only by the population of the earth.

"Special types of distribution networks, new forms of stagecraft and a development of studio equipment and technique will be required. With these must come a new and greater service of broadcasting, of both sight and sound. A new world of educational and cultural opportunities will be opened to the home. New forms of artistry will be encouraged and developed. Variety and more variety will be the demand of the day. The ear might be content with the oft-repeated song; the eye would be impatient with the twice-repeated scene. The service will demand, therefore, a constant succession of personalities, a vast array of talent, a tremendous store of material, a great variety of scene and background."

MUSIC DOES NOT NEED SIGHT.—Leaders in music attest to the fact that radio has given the masses a new appreciation of music. Some are wondering what the effect of television will be on the vast audience that has learned to enjoy good music without seeing the musicians and conductors.

Willem Van Hoogstraten, conductor of the New York Philharmonic Symphony Orchestra, was asked at a dinner in New York if he expects television to aid in the development of appreciation for classical and symphonic music that is broadcast.

"My answer is no," said Van Hoogstraten. "Television may revolutionize other forms of radio entertainment but it cannot be expected to create a devoted interest in the higher forms of musical composition because sight of the artists and their instruments invariably dulls the appreciation of the sounds created, even if the auditor is highly skilled in the art of listening.

"The ears alone should be permitted to generate one's impression. When a person sees the musical performer sight usually takes away something from his ability to hear and assimilate the tonal shadings of good music, which, after all, is something for the ears. A higher plane of musical appreciation by listeners in general should be reached without seeing by radio, even assuming that television were perfect. All this is my personal point of view. I can understand that others may want to see the artists. However, the listener who sees not but hears completely, although the players or their images are before his eyes, gets the most out of the musical score. Television in itself is a great mechanical-electrical achievement but I cannot see it as an aid in the true appreciation of music.

"Appreciation of music is the great thing to be sought," continued Van Hoogstraten. "But first one must learn how to listen. Those attending a concert should not depend too much on the eyes. Do not look at the conductor and the musicians, or the individual artists on the stage, and you will begin to hear things never heard before. If one must hear and not see, to get the most out of music, is not broadcasting of sound the ideal medium of conveying the works of the great masters to everyone?

"I realize that many people like to see orchestras and conductors, but the heart of the matter, I believe, is that it steals attention from the music. Conductors do not stand on the platform for people to look at; they are there to convey to the men of the orchestra that which the composer had in mind when the music was written, as conceived by the leader. I am much in favor of building halls in which the conductor and the orchestra would not be seen by the audience. The lights should be soft, and not shine in the eyes. The foreground should be of soft colors. One might go even further and in an inconspicuous way, use lighting effects which change softly and slowly to suit the mood of the music."

A BOON TO THE OPERA.—There are instances in music where television will be a boon. Take for example, the opera.

The broadcasters agree that they can do justice only to certain arias and acts. That is why they are not anxious to broadcast complete opera performances from the stage. The audience must see as well as hear opera to enjoy it thoroughly. That is why television is expected to stir a renewed interest in appreciation of this class of music.

"One must see and hear the opera to get the fullest appreciation," said Rosa Ponselle, soprano of the Metropolitan Opera Company. "I believe we are rapidly approaching the day when radio and the opera will be entirely reconciled by the addition of television to sound programs. When that comes it will be a great day for operatic appreciation, but I am uncertain as to whether such broadcasting will keep people away from the seats before the footlights or cause them to gather in greater numbers. We shall see. It seems that radio is awaiting television to give the theatrical part of opera the wings now enjoyed by sound."

TELEVISION IN POLITICS.—Traveling presidential candidates may be rare by 1940. The day is likely to come when they will make personal appearances before the voters by television. But there may be a danger lurking in those screen appearances if the radio waves carry them beyond the Mississippi.

Hughes toured the West in 1916. More than half the states west of the Mississippi voted against him. The sages say that he would have been President had he remained in the East. In 1884, Grover Cleveland chose to be a mystery man, so far as the great open spaces were concerned. He stayed in the East. He won. In 1910, William Howard Taft delivered what was called a poor tariff speech in Minnesota. That state and others surrounding it went strongly Democratic. In 1919, Woodrow Wilson went into the West to champion the League of Nations. He collapsed in Colorado and returned to the national capital broken in health. President Harding made a western tour and died in San Francisco.

Some of the wise men say that Alfred E. Smith should have remained in the East. But he went west in person, as candidates may do by television in years to come. Smith lost. The brown derby did not charm the West.

The Democrats said it was absolutely necessary for Smith to travel. He had been a home-staying Governor, scarcely known by sight outside the Empire State. He had to show himself to his countrymen. Hoover, on the other hand, was a national figure. California did not see Al Smith but radio lifted his words of political wisdom over the Rocky Mountains and spread them up and down the Pacific coast while the bands played "The Sidewalks of New York." He made personal appearances and speeches in Omaha, Oklahoma City, Helena, Minneapolis, Milwaukee and Rochester. Ten years ago only the people in those cities attending the political mass meetings would have heard him. But in 1928 the nation tuned in.

Hoover spoke in New Jersey and they heard him in California. He fired the opening gun of his campaign at Palo Alto and was heard in Maine. Radio in 1928 made the presidential race a national affair within the home circle and took it away from the front porch.

Possibly when television enters the campaign, red fire and bunting will come back. Gestures will be in order. The campaign will be more realistic than a mere radio battle of words. But the election bulletin boards in front of the newspaper offices are likely to disappear as the returns are flashed on television screens.

When the 1928 campaign began it was predicted that the contest would be won or lost on the radio. The man with the radio voice would win. Hoover, a shy speaker, found the microphone a friend indeed. Al Smith, at home with any audience, was hindered by the prepared speech. He could not get his personality over the microphone until he cast aside the typewritten sheets and spoke extemporaneously. Then he was at home on the air. Hoover was called a "textbook" type of speaker, while Smith was likened unto a novel.

How Broadcasting Will Change.—It is natural that the universal question "What effect will television have on sound broadcasting?" is on the tongue of all who are interested in it whether from an artistic or a financial standpoint.

The entertainers have reason to wonder what they must do to adapt themselves to the new medium. The station owners wonder what will become of the millions they have invested in equipment. The listeners wonder if their receivers will be obsolete. The answer is that television is destined to revolutionize the art of broadcasting.

All stage stars are not screen successes. Stars in the silent cinema often ceased to twinkle when the talkies were introduced. New requirements had to be met. New talent was discovered. So it will be in television. All broadcast stars should not expect to win new triumphs when radio is given eyes that enable the audience to see. They may be delightful and serene on the wings of sound alone but that does not mean that the eye will be pleased with them too. Some will captivate both eye and ear. They will be the stars of television. The entertainer who can please the eye need not worry so much about the ear. But the one who can please the ear and not the eye had better watch out.

Generally speaking, humorists have failed in broadcasting. Their jokes have usually fallen flat. The comedian needs to be seen in action. His antics and gestures, facial expressions and make-up do more to put him across. Television will give the comedian a new day—more opportunities and a greater audience than radio ever did with sound alone. Radio vision will help the world to have more laughs. The clowns will have their inning.

In the beginning probably the regular broadcasting stations will be employed to handle the sound part of the program while short waves carry the images. Eventually, the sound portion of the program may be moved to the ultra-short wave realm, too. That would mean new equipment.

BETTER SAFE THAN SORRY.—Right here is one reason why television cannot be developed overnight. Let us suppose that the regular broadcast band from 200 to 550 meters is to be utilized for sound, and short waves for the scenery. It means two transmitters and two receiving sets. As research progresses it is discovered that the ultra-short waves are ideal for both sound and sight, more economical and more efficient from a scientific standpoint. Then the transmitting stations would have to be scrapped and so would all the receiving sets.

If there were 600 transmitters and 10,000,000 television receiving sets the manufacturers would think awhile before disturbing such a vast investment. So it is far the best to determine in the beginning what waves are most suited for television and build on that foundation rather than develop on an uncertain basis and later be afraid to change to a more efficient system because of the tremendous investment that would be disturbed.

A Washingtonian once remarked to a Chicago station owner, one of the pioneers in that city, that he could have bought a 500-watt station in Indiana six years ago for $1,000 and could have made "a barrel of money."

"I wonder if you would have," remarked the Chicago broadcaster. "We rushed in here in the beginning and it didn't cost much to buy the equipment. But we have spent thousands and thousands of dollars since in development and replacement of antique apparatus. We have made money but would be far ahead of the game if we had waited until broadcasting became stabilized and then bought a station. It would have cost less. So you cannot look at this broadcasting business from an original cost basis. That is why many of the pioneers are forgotten. The man who takes them over after the development is more or less complete makes the money. We have learned our lesson in this rapidly changing field of radio. We are watchfully waiting for television. We will let the others rush in at the start, and when we think it is ready we will enter the race, fresh and ready to profit by their mistakes and experiences."

AYLESWORTH LOOKS AHEAD.—It was M. H. Aylesworth who pointed to the immediate application of television as the visual presentation of the broadcast artist. He contends that the public may look forward to an early television supplement to the regular sound broadcast programs, in which speaker, singer or musician will appear on the home television screen as purely optional features. In other words, he believes that the sound program will be received in the same manner as it is today. However, if the home be equipped with a television receiver, it will be possible to tune in the animated portrait of the performer. That this feature will prove highly attractive, no one will deny, especially in the instances of an entertainer whose personality is firmly established in the hearts of the present "blind" audience.

There are certain to be new uses for television. Just as the artistic mind of the past has capitalized on the limitations of silent movies and blind broadcasting, so will the artistic mind evolve an entirely new mode of expression to be handled by the television vehicle. It is already predicted that a variation of the futuristic art, with its symbolic abbreviation whereby a few lines and masses effectively tell an intricate story, may come to the aid of television in its early stages when detail must be sparingly used.

"If television continues for some time as a supplement to existing sound broadcasts, there will be no serious artistic problems," Aylesworth remarked. "It can strive towards better detail all the while, with laboratory progress introduced into everyday practice from time to time. Ultimately, when it possesses sufficient detail for a presentation quite on a par with sound broadcasting, it may insist on its own way in many broadcasts, with television as the main issue and sound as the supplement. That day will come when television is capable of sending complete scenes over the air. Sporting events, parades, news events, ceremonies, plays, pageants—such subjects may eventually be handled by television both in the studio and out in the field at which time the pictorial presentation will surpass the sound presentation in importance.

"While cartoonists and columnists have worked up a considerable apprehension over the effect of sight broadcasting on the present sound entertainers, I can assure you there is no great cause for worry. Although it is true that our present broadcast performers are judged purely on their microphone personalities, the fact remains that most of them have equally attractive visible personalities. Indeed much has been said regarding the present blindness of broadcasting, yet our studios are anything but blind. Many performances are attended by visitors, so that entertainers must be considerate of their visual as well as their microphone personalities. If anything, the majority of the radio artists would welcome the television feature at this time, inasmuch as it would give them a desired opportunity to display their histrionic wares.

"What television may mean to broadcast performers is perhaps best expressed in the personal appearance of the artists. Time and again the audiences before whom they appear are delighted to see them in person and carry away a permanent impression which supplements subsequent listening in, making for greater appreciation and enjoyment."

TELEVISION'S RELATION TO PRINT.—Radio is not a substitute for print. Eleven years of broadcasting from sunrise to sunset has proved that. Radio offers no serious threat to newspapers or magazines. They are entirely different. They perform different functions. One appeals to the brain through the ear, the other through the eye. But what about television that appeals to both eye and ear? How will it treat the printing press?

"Radio occupies only a minor place in the advertising world, and the newspapers should not fear its competition," said Bruce Barton, at a meeting of the Newspaper Advertising Executives Association, Inc. "Radio will never take the place of the newspaper. Television, radio and all other such devices will never replace print. For this the reasons are based on human physiology and human psychology—in other words, on human nature.

"Psychologists tell us that men and women receive 80 per cent of their impressions through the eyes, and only 20 per cent through the ears and other media. People are four-fifths eye-minded and only one-fifth ear-minded. If I were a newspaper publisher I would fear a great many things, a great many other forms of competition, before I would fear radio. Although radio occupies a real integral place in the advertising field, its place is strictly minor and limited. Radio, and even television, can never achieve the permanence of the printed page. It comes and goes with the speed of light. Hence its impression must be to some degree transitory and ephemeral. Not so with a message in print.

"In another way radio lacks the stability of print. It constitutes no record. It is, from its nature, less dependable. When I pick up my newspaper I know just where to find, for instance, stock market quotations and financial news, and just how they will appear. I infinitely prefer to read the quotations in a newspaper rather than hear them over the radio. I can study and digest them in a newspaper. Over the radio, practically speaking, I can't."

Television is a fleeting medium as is sound broadcasting. Print leaves a permanent record. It can be read at will. Television must be seen at the definite time it is on the air.

Nevertheless, it will be an advertising medium more effective than sound broadcasting because a picture is likely to leave a more indelible impression on the mind than do words uttered by an invisible person. Printed advertisements will probably be handled by television. Bold face type can be seen by electric eyes that send it through the air. An advertiser, especially during the daytime, may devote several minutes of his program to an attractively printed and interesting announcement. Housewives could read and digest it more carefully on the screen than they can the words of an announcer. And television offers opportunities for actual demonstrations.

ACROBATS INSTEAD OF NEWS.—Television, because it advertises by sound and sight, looms as a greater competitor to the press than does sound broadcasting. The fact that it handles sight, which broadcasting lacks, gives it an added weapon in its fight for supremacy in the world of business. If a television screen can carry a printed advertisement and attract millions to read it by presenting it as part of an entertainment, it becomes a competitor of print. It may divert revenue from the press and from magazines. The newspaper offers news to attract readers to the advertisements. Television will offer musical entertainment, comedy, drama, news events, dancers and acrobats, boxers and magicians to attract observers to its advertisements. A newspaper knows how much circulation it can offer. If there are 10,000,000 television receivers in the United States by 1942, as predicted, a coast-to-coast television system could offer quite a circulation—but one difficult to guarantee.

However, looking at it from all angles, there seems to be little doubt that broadcasting of sight-sound programs eventually—probably after 1940—will enter the advertising field as a greater competitor of print than radio broadcasting ever threatened to be. Facsimile transmission may some day send printed matter right into the home.

"We saw radio coming as a competitor," said the publisher of an Iowa paper. "We felt we should grasp it and link it with the press. So we ran the paper and the station together. Our revenue and circulation jumped more in two years than in the previous forty. The circulation of the paper became greater than the population of the town."

A BIRD's-EYE VIEW FOR ALL.—Turning to aviation, numerous possibilities are seen for television. The bird's-eye view will take on a new meaning. An electric eye linked with a radio camera from a lofty perch can photograph a scene and flash it to earth by short waves. On the ground it will be intercepted, recorded on a film and rebroadcast throughout the country by television. Californians may see what the aviator over Manhattan sees as he flies above the skyscrapers. The entire nation may be taken for a television tour across the Grand Canyon, down the St. Lawrence or see Broadway's Great White Way converted into a nocturnal fairyland of electrical glow. Spectators at home may see how their city looks when scrutinized from the sky. These are peaceful rôles for radio vision.

War offers new opportunities. Planes equipped with television transmitters will look down on the enemy and broadcast the scene. Televised maps will be flashed to the planes. Those directing the battle from behind the lines will see exactly what is going on up front. The gunners will see if they make a bull's-eye, how the barrage is falling and where to direct the shots. Aircraft over the sea may make radio photographs of convoys and fleets, then broadcast them by television to submarines lurking beneath the surface ready to attack. Radio waves are audible under the water as well as under the ground, and so television can dip faces through the sea.

Police are likely to find television a valuable assistant in war on crime. Pictures of criminals, fingerprints, photographs of missing persons will be televised for reception by patrol cars and police booths. Printed orders from headquarters will be flashed on screens. Officers and patrolmen on the line of duty will look in on the police line-up at the station house.

THE DRAMA OF EXPLORATION.—Adventure and exploration linked with television cameras give the imagination an opportunity to function. It will be recalled that the members of the Byrd Antarctic Expedition were thrilled by familiar voices broadcast special to them from Pittsburgh and Schenectady. Out of the darkness of the long winter night, through various climes and a mixture of weather, came voices the identities of which were faithfully preserved, and the ring of the voice was true, despite the long flight across land and sea, across jungles and mountains, across the Tropic of Cancer, the Equator and the Tropic of Capricorn, finally to strike a slender target of copper antenna wire stretched between two masts reaching up from the ice.

If all that is possible—and it has been done—why should radio not carry sight to and from the far distant points of the earth? Isn't it feasible to believe that some day an explorer will soar over the South Pole with a television camera, just as the plane *Floyd Bennett* carried a motion picture eye? Then, instead of waiting for a ship to bring the films to civilization and to theaters throughout the land, radio will flash the scene around the globe so that many millions will see exactly what the aviator views and at the instant he is seeing it. The fact that messages have traveled back and forth from the isolated regions leads those who have faith in science to believe that as the ear hears so shall the eye see.

NO SUBSTITUTE FOR TEACHERS.—The school children will see history made when the television screen is hung alongside the blackboard. It will not dispense with the instructor who covers the three R's. Television will merely supplement teacher, books and chalk. The motion picture, phonograph and radio are merely additional tools for the teacher. Such is the rôle of television. It will carry historic scenes that books will describe for students who go to school in later days. Television will be a timely instructor.

The National Advisory Council on Radio in Education at its first assembly, May 21-23, 1931, in New York, declared in its report that television broadcasting is in an advanced experimental condition.

"It has not yet been possible to establish transmitting stations capable of giving reliable television service over a considerable area, nor to provide on a commercial scale receivers which give a clear, bright picture of an acceptable color, adequate detail, satisfactory size, freedom from flicker, of sufficiently wide angle of view, and of the requisite steadiness of position," stated the Engineering Committee. "The problems involved are under active investigation, and there is a likelihood that within the next few years equipment of this sort will be commercially available and that at least a moderate number of television broadcasting stations capable of supplying program material to those having suitable receiving equipment will be in operation.

"The problem of network syndication of television programs is in a less advanced state. If a program for television transmission is recorded on a motion picture film, methods analogous to the electrical transcription will doubtless become suitable for syndication. It is also possible that wire line facilities capable of carrying television programs will be developed, although these do not exist even experimentally at this time."

It is believed that the value of television for educational purposes will be largely dependent upon the amount of detail which the picture can carry. If the development of television during the next few years leads to pictures of such detail that lecture room demonstrations can be readily and clearly reproduced, and if some impression of the personality of the lecturer can be gained by the observer, and if the range of transmission and reception is such that large groups of people can successfully receive lectures and demonstrations, it is anticipated that television may have a substantial educational value and a wide application.

It should be pointed out, according to the Advisory Council, that much confusion exists among the public as to the exact meaning of the term television. Comparatively blurred, dim, flickering and unsteady images, carrying little detail and simultaneously visible to only one or two persons at a given receiver, and then only in a darkened room, are claimed by some to constitute successful television. Equipment capable of yielding such limited results is on the market to a slight extent, but is obviously of no significance to educators. From the point of view of the educator, a picture of an entirely different and greatly superior character is strictly necessary. Any educational project based upon pictures which do not meet reasonably high specifications will find the application of television a handicap rather than an assistance, inasmuch as a poor picture is rather a distraction than an instructional agency.

Television is at present in such a state that in general the mode of transmission and its romantic interest attract a major portion of the observer's attention. In consequence it parallels the condition of radio broadcasting at the time when the quality of transmission was at a level where criticism of the program or of the artists was not practical, according to the report. The medium of transmission was not sufficiently precise or constant in its action to enable criticism to be well founded. Until television reaches a stage where the mechanics will be forgotten and attention concentrated on the program itself, its utility in education will be small.

IT ALL TAKES TIME.—It seems that there are so many fields in which television can function that the work may never be finished. It will require long years to take advantage of all it offers. No one should expect that by 1940 television will be perfect and performing all that it eventually will perform. By that time it will be just getting under way. Televised preachers and pulpits, televised fashion shows from Paris, televised Passion Plays from Germany, televised Niagara are not for the morrow. The greatest miracles in seeing by radio belong to the future. It was a far cry from the candle to the electric light, from the schooner to the ocean liner, from the horse to the 8-cylinder motor car, from the balloon to the 400-mile an hour airplane, from the stereopticon to the talking motion picture, from Faraday's dynamo to the powerful generators that harness the cataract of Niagara. Many years intervened between these stages of progress so man must be patient while radio broadcasting leads on to television.

The past hundred years might well be called the Century of Electricity. Those who have doubts about television's future will do well to reflect on all that has happened.

Little did the world realize in 1831 that Michael Faraday's many discoveries in physics and chemistry—especially his great triumph, the generation of electricity by causing magnets and coils of wire to rotate relatively to each other—marked the beginning of a century of electrical wonders.

Faraday, so engrossed in his research and so close to it, probably little dreamed that the day was to come when networks of wires would be crisscrossed on poles over the land and through conduits under the ground carrying electricity for lighting, to perform useful work in the home, factory, office and on the streets. Who in those days dreamed of a world-wide communication system with telegraph, telephone, radio and television all supplementing each other to annihilate the distance that separates city from farm and nation from nation.

Today these electrical wonders are accepted as a matter

of course. The thrill and novelty of hearing music that travels across the Atlantic without the use of wires are no longer a front page story. In the news it is no different from New York hearing music performed in Newark across the Hudson.

Each year will bring new discoveries in the electrical field. There is much ahead. Television, which some frown upon today because it is not clear, because the image is small and streaked, will surprise its most optimistic followers and prophets. What appears a fantasy today is likely in years to come to surpass even the dreams of this ingenious age.

It is no wonder that television has its skeptics. It is a miraculous science, almost unbelievable. It is uncanny when a man can smile in London and that smile is seen instantaneously 3,000 miles across the sea with no connecting link except an invisible medium.

The great scientist Tyndall found it difficult to follow Faraday. It seemed to him that "there was a vast vagueness of immeasurable hopefulness in Faraday's views of matter and force." Yet he conceded that the Faraday "discovery of magneto-electricity is the greatest experimental result ever obtained by an investigator. It is the Mont Blanc of Faraday's own achievements."

Faraday had built "an electrical machine." It was a crude affair compared with the great generators of today but it did the trick. There was a disk of copper between the poles of a large magnet. A metal brush rested on the shaft of the disk and another on the rim. Then he turned the disk and the result was a continuous current—the first dynamo. Mechanical motion had been converted into electrical energy. No longer was a battery the only source of a steady electrical current. The life-blood of industry—electricity—was beginning to flow in copper wires that would some day girdle the globe to work for mankind.

That was the beginning. The march of electrical progress since that day has mustered a multitude of men to do their bit for the evolution of television—the era of which has dawned. But this monarch of radio, the Mont Blanc of television, is still in the distance!

Chapter Fifteen

FACES AND SCENES ADRIFT

A DRAMA facilitated by science is unfolding, the ethereal curtain is slowly rising, television in later acts is going to telescope the world bringing scenes of grandeur to millions of screens in a performance P. T. Barnum no doubt would dub the "greatest show on earth."

Television might well be called an electrical palette upon which art and science join to blend faces and scenes and then the touch of an electron brush sends them adrift in the emptiness of space. It is the magic that transforms pictures of people and places from light into electricity, from electricity that flows on a wire to radio that spreads through the air. But in the twinkling of an eye the ethereal phantoms are all changed back into a festal of light as if touched by some master showman's wand that reaches stealthily down from the sky.

Some day, in years to come, the American family at home in any metropolis, town or hamlet may watch a Roman pageant in that Eternal City beyond the Alps or view a military tattoo under the glare of an Indian sun. It will all depend upon the wave length selected. Berlin will be just a hair's breadth away from Montreal. A finger will turn the dial that separates Tokyo from Budapest. The slight turn of a knob and the scene will shift from Africa's jungles to the land of the midnight sun. And these pictures will speed around the globe in a split second as years come and go. Television promises a vivid spectacle.

Man marveled when the buzz of a bumble bee crossed the continent; when the song of the nightingale in England was heard in America; when the whisper of Ezra Meeker's voice feeble with age was audible across the nation many years after he had helped to blaze the Oregon Trail in a covered wagon. Radio did that. Now television is thrilling the world. And with sufficient electrical power to hurl it into the infinite, a wink, a frown or a smile may some day girdle the earthly sphere seven and one-half times in the tick of a clock. Radio travels at the speed of light covering 186,000 miles in a second.

A WORLD-WIDE MIRAGE.—The time is likely to come when short waves will empower spectators in Europe, Asia, Africa and the Orient to see Niagara adrift in space above the hemispheres. It will be a world-wide mirage on the television screens. Japan and Germany, England and Australia, Brazil and the Argentine will see the tons of turbulent water tumble over the precipice into the gorge below while loudspeakers reproduce the thunderous roar of the tossing torrent. The Japanese schoolboy will catch a glimpse of the falls almost as vividly as the honeymooners at Prospect Point.

New Yorkers will turn a dial that begins a whirl and swirl of imperfect vagueness, which upon refined tuning takes them to Egypt, the land of early dynasties, to Luxor and to Karnak where decorative obelisks, their sides carved deep with hieroglyphic inscriptions, stand as solitary monuments to fallen cities and to kings long dead. There they will see the Great Sphinx at Gizeh, crouched in the sands for centuries little suspecting that some day his weather-beaten countenance would be televised in a radio drama.

Londoners may set their dials in tune with an invisible ray that prolongs the range of their optic nerve across the Atlantic and into the great Far West to Glacier National Park. There they will catch a magnificent sight of the snow-capped Rockies, and perchance Old Faithful geyser belching steam and spray into the sky. The globe will be a great kaleidoscope.

Geography will be an animated subject as a flutter of light brings pictures of people and the scenery of nations to school and home just as radio brings voices and music from far and wide. The grand old Mississippi, the Danube and the Rhine will flow on television screens while orchestras broadcast "Ole Man River," "the Blue Danube" and the "Ride of the Valkyries."

Television like a graphic roll will spin its scenes around the globe. It seems destined to turn the entire universe into a vast cosmopolitan theater in which many millions are seated to enjoy a production of world-wide magnitude. No international language will be needed. The pictures, smiles, acting and laughter, no matter from what amphitheater they come, from Moscow or Shanghai, from San Francisco or Paris, will fascinate all without the use of words. The motion picture will tell the story in picturesque detail portrayed in black and white. Commercial television in tints and natural hues belongs to the distant future.

Some of these things seem fantastic, but there is a magnificent vagueness about them. It was difficult for many minds in 1900 to comprehend that human thoughts could span the ocean without the use of wires linking the distant shores. That was only thirty-odd years ago. And fourteen years elapsed from the day of Marconi's transatlantic triumph until the spoken word found its way through the air from Washington to Paris. Eleven more years went by before the image of a face was tossed from London to New York. Scientific progress requires time and patience. Television is part of the miraculous scheme—so spectacular that a year is but a fleeting moment in the gigantic task of making it practical for everybody.

THE EYE IS FICKLE.—It is one thing to please the ear and a vastly different task to please the eye. The eye is quick to reject by a drop of the lid or a turn of the head. The ear is not so equipped for censorship. To be perfect television must emulate the talking cinema else the fickle eye may close its lid. That introduces a challenge to the research corps. They can project a motion picture from the rear of a theater to the screen on the stage. But they must master new forces, acquire new skill and technique before they can project Eiffel Tower or Gibraltar over the sea. Once they can successfully televise a mouse it will be easy to send an image of the elephant.

Historic events will be preserved for posterity on television films. There will be libraries of talking pictures, miles and miles of film that portray scenes of war and peace. Years afterward the reels will be taken from the fireproof archives to recall memories of the past as radio gives the images of characters long dead, the power to live again, to walk and to talk on silver screens. Television is a parallel to the sound motion picture with its drama, tragedy, comedy and newsreel.

To reveal the thrilling possibilities let us go back a bit, into the land of what-might-have-been, where the imagination links the future with the past. Charles A. Lindbergh is at Curtiss Field ready to begin his famous flight to Paris. The electric cameras with their all-seeing eyes and sensitive lenses are on duty to snap the scene.

On May 20, 1927

It is 2 o'clock in the morning. Television screens depict a murky, dreary scene. Puddles caused by a thunder storm's deluge around midnight glare like tiny lakes in the flood of the electric lights that are now illuminating the television scene. The silver nose of the *Spirit of St. Louis* glistens through the hangar door upon which a television camera is focused. Word has been broadcast that Lindy is preparing to go. Crowds are collecting. Automobiles line the roads around the field and on the television screen the Nassau

FACES AND SCENES ADRIFT

county police are seen rushing about to keep the mob off the flying field.

Wisps of fog blow across the field and the televisors catch them. It is a dismal scene but nevertheless a dramatic one for the millions who are forsaking slumber to watch an historic event. They hear the chatter and comment of the crowds. The shouts of the police. Complaints of the wet field. Microphones alongside the radio camera pick up words that come from the crowd. Some one says it is the height of folly for any plane to take off, even on a short flight. A spectator at the field looks skyward and holds out the palm of his hand. He says it is raining again. But some one runs out of the hangar and reports the sky to the north is clearing. Apparently the threatening conditions are only local.

A policeman's motorcycle roars across the television screen. He is clearing a path through the crowd for the pilot. It has stopped raining. The big doors of the hangar swing open revealing a graceful plane. A truck backs up to the doorway. The *Spirit of St. Louis* is turned around. The tail is lifted up by careful hands and is made secure with ropes. Every precaution is taken to avoid strain before it gets into the air. It is bound on a long journey on an uncharted pathway along the Great Circle Route that links two continents.

The television screens of all America are illuminated with activity. Motorcycle policemen are seen to surround the truck and the silver bird is pulled ignominiously along tail first across the field as a corps of television photographers follow with their electric eyes and microphones. Mechanics are seen to stoop occasionally as they walk along to feel the wheel bearings for fear they might heat up under the load of 200 gallons of fuel already pumped into the tanks.

It is 5 o'clock. The rain is sprinkling. The truck moves slowly toward the runway so that the plane will ride tenderly over the rough spots and puddles that dot the field. The grass is wet and the ground soggy.

THE SUN COMES UP TO HELP.—The television scene is becoming a bit clearer. The clouds in the east are breaking and the first faint streaks of light appear. Soon the radio men will not need the artificial spotlights and flares to illuminate the scene. At last the nose of the *Spirit of St. Louis* with a canvas cap over the motor is at the head of the runway.

The ship is facing the rising sun. A closed car approaches. A youth in army breeches and a tight woolen sweater steps out. Men standing on the nose of the plane are pouring in the gasoline. Fellow aviators realize this flier will soon be off and they are shaking hands and wishing him good luck. The spectators on the field and at the television sets are excited and anxious. Those at the field have wet feet. Those watching by television are comfortable at home, many of them in night clothes. They wonder if there is any bottom to those hungry gasoline tanks. Finally they see the men climbing down from the plane. They hear a mechanic tell the flier that there are 451 gallons in the tanks, 150 more than the plane ever lifted. It is a dramatic moment.

A mechanic turns over the motor. There is a terrific roar on the television screens as the associated loudspeakers reproduce the noise. The birdman is seen donning his fur-lined flying suit. His helmet is shoved back and the goggles rest high on the forehead. He gazes off into space. He is the most unperturbed man on the field. He climbs into the pilot's seat and warms up the motor. The throttle is open and the great man-made bird roars and flutters. The television eyes are not missing a thing.

Some one runs up excitedly and asks him if he has forgotten his rations.

The microphones on duty pick up his answer. The electric cameras point at him.

FACES AND SCENES ADRIFT

"I have five sandwiches. That's enough. If I get to Paris I won't need any more, and if I don't get to Paris I won't need any more, either."

"How is it?" asks the pilot.

"She sounds good to me," replies the mechanic.

"Well, then I might as well go."

It is exactly 7:52 A.M.

"So long" he calls from the tiny window of the plane as he waves to the crowd.

MOMENTS OF EXCITEMENT.—The blocks are pulled from beneath the wheels. The motor roars. Television eyes located all down the field watch the heavily burdened plane lurch slowly down the runway. The wheels find it difficult to travel over the bumps and soggy field. She does not seem to get up flying speed, at least not enough to rise with the load. The television spectators groan. So do those at the field. The plane looks nose heavy and as if it might plunge over on its nose at any instant. It must lift quickly or strike a gully at the end of the runway. Suddenly it hits a bump which throws it upward on the television panorama. But the wheels come back to earth. It has not enough flying speed. It looks as if the craft is too heavy. Suddenly as if some unseen force were lifting the wings they leave the ground and the plane just skims over a tractor which is directly in its path and near the last television camera.

The camera man turns his lens toward the plane. It is seen to clear the electric wires by barely twenty feet. But it doesn't seem to rise high. There are trees ahead. Lindbergh apparently sees them through his periscope. He turns a little to the right and selects the point where the foliage is lowest. The silver wings sweep by and the machine begins to climb. The sun creeps out from behind the clouds and smiles on the *Spirit of St. Louis* as it dips off over the horizon. It is just a mere speck in the television picture now.

A fleeting bird is flying across Long Island Sound on a

course that leads to Rhode Island and over Massachusetts Bay to Nova Scotia and on to Paris, 3,610 miles away. Lindy is out of sight. It is time for the television audience in the East to go to work. Californians may catch a few hours' more sleep. The next scene will come from Le Bourget flying field in France. It is almost midnight in Paris 33½ hours later, when the landing lights flash across the sky and 100,000 pairs of eyes are on the watch for Lindy.

How could it be done? How could the earth's population be spectators at such history making events? Television is the answer. It is the wizardry of the age.

SCIENCE BECKONS TO MAN.—All the intricacies of television will not be completed in this century. The scientists of other generations will pick up where the contemporaries leave off. Television is too preponderant a work for an individual. It is too great a task for a generation of mathematicians, optical experts and radio engineers. Only time measured by years will solve the many problems that eventually will open the way toward the goal of ultimate perfection. No one man will wear television's crown of success. New electrical and optical instruments, new vacuum tubes and "cold" filamentless bulbs, new cameras, new microphones undreamed of today will be discovered in the march of progress. And in the end television will be so simple!

It is well to remember that man was toiling on the transmission of pictures back in 1840 and he will be toiling on and on with television, ever trying to improve it, long after 1940 passes by the milestones and down the dim corridors of time.

The television day has dawned. But it is a long time between the sunrise and the sunset of a new science. It is measured by centuries in which the span of a human life is but a fleeting moment. In that time many men play many parts. They contribute their mite to the miracle called television.

FACES AND SCENES ADRIFT

Men, my brothers, men the workers,
    ever reaping something new;
That which they have done but earnest
    of the things that they shall do.

For I dipped into the future, far as
    human eye could see,
Saw the Vision of the world, and all
    the wonder that would be;

Saw the heavens fill with commerce,
    argosies of magic sails,
Pilots of the purple twilight, dropping
    down with costly bales:
    .   .   .

Eye, to which all order festers, all
    things here are out of joint.
Science moves, but slowly, slowly, creeping
    on from point to point.
                "Locksley Hall," Tennyson.

EPILOGUE

Television is envisioned as a boon to many activities in human life. In it lurks the germ of entirely new international relationships. It looms as a revolutionary social force. It threatens radical changes that will speed the tempo of a slow-pulsing industrial world, the wheels of which are stopped or moved irregularly in the throes of a business depression. Television may be the hero of the hour as it emerges from the research laboratories to answer the call of a new era.

Prominent men, leaders in various fields including science, education, religion, drama, politics, business and warfare have been invited to participate in a symposium, projecting their minds into the future. This is what they foresee ten to twenty years ahead.

BY BRUCE BARTON

BATTEN, BARTON, DURSTINE & OSBORNE, INC.

Inevitably, television will have an important influence on advertising, newspapers and magazines, but it seems to me unlikely that it will displace either newspapers or magazines. Existing media of communication and education are modified by new developments but are not usually displaced by them.

I recall that in the early days of the telephone the telephone company advertised: "Don't travel. Telephone." This appeal was quite promptly withdrawn because they discovered that the more people traveled the more they telephoned. It seems to me likely that the more the public of the future is informed and educated by television the more, rather than less, it will appreciate and depend on the newspaper and the magazine.

The thrilling thing about the universe is that everything has to change. In our advertising agency we try to remember this and to keep alert. We were one of the first of the agencies to recognize radio was destined to play a large part in advertising. Our radio business today runs into many millions of dollars, yet it has been built up without diminishing our newspaper and magazine appropriations.

BY REAR ADMIRAL RICHARD E. BYRD

After Peary discovered the North Pole on April 6, 1909, and Amundsen found the South Pole on December 14, 1911, many

EPILOGUE

days passed before they completed the hazardous trek back across the ice to notify the world that the long-sought goals had been reached, and man had stood at the top and the bottom of the globe.

When Floyd Bennett and I flew over the North Pole on May 9, 1926, and when Balchen, McKinley, June and myself encircled the South Pole on November 29, 1929, we carried radio apparatus. It was only a few hours before newspapers were on the streets of New York and other cities telling the story of our airplane conquests.

Radio not only carried the bulletins of success out of the desolate regions, but it brought us messages and news from home. We picked up music and voices that relieved the monotony of Antarctica. We were only one-twentieth of a second from New York by radio, although 9,000 miles away.

It is of interest to explorers to know that the next step in radio science may equip them with television. It would be miraculous, if in years to come a group of adventurers visited Little America and by turning dials saw faces of friends smile at them through a blizzard raging outside. And in exchange the explorers might project the Antarctic scene with the penguins as the actors. Then the folks at home would catch a glimpse of a remote section of the earth seldom seen by human eyes.

BY DR. GEORGE B. CUTTEN

President, Colgate University

He is a brave man who will dare to prognosticate what even the immediate future has in store for us as a result of the marvelous development of science, the threshold of which we are just crossing. One of the subjects which promises most is television. The moving picture, with the accompanying dialogue or description, has been a great step in advance; but the television will be to that even more of an advance than a telephone conversation is to a letter.

One cannot refrain from thinking what it will mean in education where the most noted lecturer in the country will not be confined to a single classroom or university, but while lecturing to his immediate audience he may be seen and heard in every other university in the world. Nor need this reception be confined to university classrooms, but a college education may be available to every person in his own home. The question really comes whether in the future colleges as formal institutions will be necessary, and if the attendance of classes in any one place will not become as obsolete as the buggy of twenty-five years ago.

While we think particularly of classroom lectures, with the improvement and development of television, there seems to be no part of college work that would not be immediately available to every individual in his home. Starting in with classwork in the morning, one might see a football game in the afternoon, and attend a college play in the evening, without moving out of his chair.

Just where this will all lead is a secret which the future is as yet guarding carefully.

BY DR. LEE DE FOREST

Within ten years television, by wire or radio, will be in the majority of homes of the well-to-do in the more thickly populated sections of America. It will, I trust, be introduced, maintained and regulated on a far wiser and more business-like basis than is our aural radio.

The television receiving equipment will never be as compact, simple and economical as are good radio receivers today. But whether supplied by ultra-short waves or over wire, we may depend on the picture being far more free from static or fading than is radio music.

One grand advantage of television over radio, for which all apartment dwellers might be devoutly thankful, should be that the neighbors' vision won't disturb others—that is, of course, if they will obligingly cut off or subdue the vocal accompaniment, which they too often won't!

Steady, clear, screen projected pictures in almost black and white, one to two feet square, will be the usual form in the home. These, while not revealing all the infinite detail of a motion picture, will have the surpassing value and fascination of being actual transmission of living personality and actual coincident events. There will always reside a compelling thrill with which no "canned" picture, however lovely, can ever successfully compete.

To the home screen, on the wall or cabinet, will then come daily and nightly scenes from distant theaters, fashion manikin parades, ball parks, boxing arenas, cruising dirigibles, river steamers—there will be no end of rich variety for the enterprising television pick-up. The horizons of all will be enlarged, home life will be far more attractive. Television should do much to restore acquaintance with home and family.

Theater producers and artists will welcome this enormous increase in their audiences. For television will be developed along reasonable, business lines, where those who use it will pay for their enjoyment, by monthly rentals, or metered wired service—small in each instance, but aggregating the large sum necessary to provide the talent and the intricate equipment involved, without objectionable recourse to the incessant advertising ballyhoo, which radio broadcasters now seem to feel so essential to their own existence.

BY THE RIGHT REVEREND JAMES E. FREEMAN
The Bishop of Washington

When Edward Bellamy wrote his book, *Looking Backward,* he projected himself into the centuries ahead and tried to envision the procession of the years. Today our prophets and seers are looking forward and, with a fine recognition of the unparalleled advances we have made along scientific and mechanical lines, are conceiving of a new age that shall far outstrip the present.

Radio has made the world a whispering gallery and has so emphasized the intimacy of our fellowship that it is tending to break down national and racial prejudices. Now television is to bring within the confines of our homes scenes and personalities far removed from us.

Some fear that in the realm of religion and corporate worship it may tend to weaken the Church and make us unmindful of the ancient admonition: "Forget not the assembling of yourselves together, as the manner of some is." Our judgment is against this view. Anything and everything that can render religion more articulate must give freshened demonstration of the value of that which the Church stands for. Radio and television must quicken the appetites of men for things spiritual. Nothing makes us yearn more for the companionship of our fellows than the suggestion conveyed to our senses of the meaning and worth of that which reaches its highest satisfaction in places where great assemblages are met.

BY MAJOR GENERAL JAMES G. HARBORD
Chairman of the Board, Radio Corporation of America

If television advances in the next twenty years as those who are watching it in its laboratory stages hope, it will change the conduct of future wars as much as giving fuller sight to a man partly blind would change the range of his activities. Assuming, for purposes of a forecast necessarily fanciful, that within a score of years television can send from a moving transmitting station an image as detailed as that on a motion picture screen today, the possibilities of its applications in war set the imagination spinning.

With such television "eyes," strengthened by telescopic lenses, aircraft flying over enemy territory may carry back to future army headquarters the view that would lie before an aircraft observer with high-powered field glasses, to supplement airplane pictures and maps. On such information plans for attacks could be made. As attacks move forward future generals may see spread before them on screens moving images of their men advancing, notice the massing of the enemy at a certain point and shift the attack quickly to a weaker spot.

It is within the realm of bad dreams or a delirium that unmanned aircraft loaded with explosives and guided by remote radio control may be sent far into enemy territory. Men at a television screen may see the country under such aircraft and select targets as accurately as if they were in the cockpits. Television equipped torpedoes may follow ships, no matter how they dodge.

Television promises eventually to take a place on a par with sound broadcasting today. In that case it will be a factor in the struggle of morale in wars to come. Every one of the millions of home radio receiving sets probably will be a target for enemy propaganda. The television watcher may see in his living room motion pictures of his nation's soldiers in foreign prison camps, for example, and—to offset that—be given convincing looks at his well-trained, well-equipped troops at the front.

When such imaginings flit through our minds it is pleasant to think that television in times of peace will take its place beside sound broadcasting as an influence toward international understanding and goodwill, and toward making war less likely.

BY COLONEL THEODORE ROOSEVELT

Governor General of the Philippines

Of course, television will have a considerable influence on politics, especially in national and state-wide campaigns. It, combined with radio, will undoubtedly cut down the number of places in which national or state candidates speak, but it will merely cut them down. It will not eliminate them. There is something which a candidate gets from appearing in the flesh before an audience, from visiting the town or city, which neither television nor radio can replace.

In this, much the same maxim holds as does on inspections. A report of conditions may be studied in an office, but an inspection in person builds morale and gives a point of view which cannot be discounted.

I believe the great mission that will be accomplished by television and radio is that whereas thirty years ago only an infinitesimal number of people in the nation ever heard the voice of a President or saw his picture outside of a few lithographs or some mediocre newspaper reproductions, now the vast majority of a country will see him as well as hear his voice. I think this will have a great effect, for it cannot but stir the nation to a lively interest in those who are directing its policies and in the policies themselves. The result will normally be that we may expect more intelligent, more concerted action from an electorate. The people will think more for themselves and less simply at the direction of local members of the political machines.

At the present, there are but a few men in the country who are known equally in Oregon, Kansas and New Hampshire. It is difficult for a man to become a national leader. With television and radio this should be greatly changed, and the number of leaders known to the general public multiplied many times.

BY S. L. ROTHAFEL (ROXY)

Projecting my thoughts into the future, how do I see television affecting the stage and screen? Will the theater pass out of existence or will it be bigger and better, because television may extend its range?

The theater will never pass out of existence. All the world's a stage and all the world has always wanted to see itself mirrored. In the history of the theater, we find periods where its hold on the people seemed to be lessening. In a few years, it regained its important position. Television, I think, in twenty years, will be the mirror, mirroring the stage. And because of television, the stage will be greater than ever. This, because of the economics. Cities will have the actual theaters and the villages the vision.

Already in our plans for Radio City, television is given a place. I am inclined to believe that in five years, because our scientists work with such amazing rapidity these days, the stage and television will be one. As to the effect, going to the question of whether the theater will pass out of existence or will be bigger and better, some evidence may be gathered from the fact that innumerable artists, practically unknown before their broadcasting, have become sensational attractions in the theater. Also, personal appearances of talking picture stars are more evidence. We are all human and we all want to see, "in the flesh," those we love.

APPENDIX

THE CALENDAR OF WIRELESS-RADIO-TELEVISION

640 B.C.—Thales of Miletus notices that amber, after being rubbed, acquires the property of attracting straws and other light objects.

1600 A.D.—William Gilbert publishes his work *De Magnete* in which he uses term "electric force."

1650—Otto von Guericke invents the air pump and first frictional electric machine.

1654—Robert Boyle observes that electric attraction takes place through a vacuum.

1666—Sir Isaac Newton performs fundamental experiments on discovery of the spectrum.

1676—Olaus Roemer discovers that light travels at a finite velocity.

1725—Stephen Gray observes that electric forces can be carried about 1,000 feet by means of a hemp thread, thus discovering electrical conduction.

1733—Dufay observes that sealing wax rubbed with cat's fur is electrified but differs from an electrified glass rod. He terms one "vitreous" and the other "resinous." Franklin later introduces terms "positive" and "negative" electricity.

1745—Musschenbroeck of Leyden discovers principle of the electrostatic condenser.

1749—Benjamin Franklin by his celebrated kite experiment proves lightning is an electrical phenomenon.

1780—Luigi Galvani makes historic observation relative to twitching frog legs, which leads to invention of voltaic cell. He calls it "animal electricity"; thus history records him as discoverer of current or "galvanic" electricity.

1794—Alessandro Volta invents the voltaic cell.

1800—William Herschel discovers infra-red rays.

1801—Humphrey Davy displays first electric carbon arc light.

1819—Hans Christian Oersted discovers magnetic action of an electric current and publishes an account of the influence of galvanic current on a magnetic needle.

1820—Johann Schweigger invents the galvanometer.

1821—André M. Ampère makes research in electricity that is re-

sponsible for relationship between electricity and magnetism.
1825—Georg Simon Ohm propounds the law named for him—Ohm's Law.
1827—Wheatstone coins term "microphone" for an acoustic device he has developed to amplify weak sounds.
1830—Joseph N. Niepce and Louis Daguerre produce first practical process of photography.
1831—Michael Faraday formulates laws of electromagnetic induction that lead to development of magneto and dynamo.
1831—Joseph Henry discovers self-induction, improves the electromagnet and makes the first electric bell.
1832—Samuel F. B. Morse discusses his idea of the telegraph.
1838—First induction coil is made by Charles Page of Washington.
1838—Steinheil discovers the use of the earth-return later utilized in telegraphy, telephony and wireless.
1847—Thomas Alva Edison born February 11 at Milan, Ohio.
1849—John Ambrose Fleming born November 29 in England.
1856—Caselli sends designs by telegraph utilizing a cylinder covered with tinfoil on which the figures are drawn in insulating compound by a contact pin or needle traveling over the cylinder.
1857—Geissler produces a vacuum tube.
1858 (Aug. 16)—First transatlantic cable is opened with exchange of greetings between President Buchanan and Queen Victoria.
1861—Philip Reis of Germany designs a make-and-break platinum contact microphone with which musical sounds but not speech are transmitted.
1865—Wilhelm Theodor Holtz builds an induction machine.
1867—James Clerk Maxwell outlines theoretically and predicts action of electromagnetic waves.
1869—Hittorf, of Münster, performs a number of experiments with tubes having comparatively high vacuum.
1870—Varley discovers that sound may be emitted from a condenser.
1872 (July 30)—First patent for a system of wireless issued in United States to Dr. Mahlon Loomis of Washington, D. C., who in 1865 made a drawing to illustrate how setting up "disturbances in the atmosphere would cause electric waves to travel through the atmosphere and ground."
1874 (April 25)—Guglielmo Marconi born at Bologna, Italy;

father, Joseph Marconi (Italian), mother, Anna Jameson (Irish).

1875—Alexander Graham Bell invents the telephone.

1875—Thomas A. Edison observes the phenomenon "etheric force."

1877—Emile Berliner in Washington observes that the resistance of a loose contact varies with pressure and he applies the principle to microphone design.

1877—Edison patents a telephone transmitter of a variable resistance amplifying type in which the resistance element is a button of solid carbon.

1878—Sir William Crookes invents Crookes tube and demonstrates cathode rays to illustrate their properties.

1878—Francis Blake designs a telephone transmitter utilizing a block of hard carbon and a vibrating diaphragm.

1878—Hughes in London designs an extremely sensitive inertia transmitter and revives the term "microphone." He discovers phenomena upon which action of the coherer depends.

1880—J. and P. Curie of France discover piezo electric effect later applied to hold radio stations on their exact waves thereby minimizing interference.

1880—Trowbridge discovers that signaling can be carried on by electric conduction through the earth or water although the terminals are not linked metallically.

1882 (March)—Professor Dolbear is awarded a United States patent for wireless apparatus. He states that "electrical communication, using this apparatus, might be established between points certainly more than one-half mile apart, but how much further I cannot say."

1883—Edison discovers what is called "the Edison effect," a phenomenon occurring in an incandescent lamp, in that an electric current can be made to pass through space from the burning filament to an adjacent cold metallic plate.

1884—Paul Nipkow of Germany invents television scanning disk.

1884—Ader of France develops a multiple carbon pencil microphone for picking up musical programs.

1885—Edison, assisted by Phelps, Gilliland and Smith develops a system of communication between railroad stations and moving trains by means of induction. No connecting wires are used. This is Edison's only patent on long-distance telegraphy without wires. (He filed the application on May 23, 1885, and the patent No. 465971 was issued December 29, 1891. The Marconi Wireless Telegraph Company purchased it in 1903.)

APPENDIX

1885—Sir William Preece in experiments at Newcastle-on-Tyne demonstrates that in two completely insulated circuits of square form, each side being 440 yards, located a quarter of a mile apart, telephonic speech can be conveyed by induction.

1886—Dolbear patents a system of wireless by means of two insulated elevated metallic plates.

1886—Professor Heinrich Hertz, a German physicist, proves experimentally that electric waves are sent through space with the speed of light by the electric discharge that takes place when a spark is made by an induction coil or a static machine.

1886—Edison applies for a patent on telephone transmitter filled with granules of hard coal.

1890—Anthony White invents the so-called solid-back transmitter.

1890—Professor Edouard Branly of France develops the coherer which considerably advances radio reception because of its properties as a detector.

1890—C. Francis Jenkins begins search for new appliances needed for success of Nipkow scanning disk.

1891—Nikola Tesla experiments with high frequency currents and discovers principle of the rotary magnetic field applying it in practical form to the induction motor.

1892—Preece signals between two points on the Bristol Channel at Lochness, Scotland, by a system that employs both induction and conduction to affect one circuit by the current flowing in the other.

1894—Rathenau signals through three miles of water by using a conductive system of wireless.

1895—William Conrad Roentgen announces discovery of X-rays from a Crookes tube excited by electricity.

1895—Marconi sends and receives his first wireless messages on his father's estate at Bologna, Italy.

1895—Marconi proves that electric waves can be transmitted through the earth, water or air by means of sparks producing high frequency electrical oscillations.

1896—Marconi files application for the first British patent on wireless telegraphy. Experiments proved his system would communicate for at least one and three-quarter miles.

1896—Marconi sends a wireless signal at Salisbury Plain, England, across a two-mile range.

1897—Marconi on tug boat receives messages from Needles on Isle of Wight, 18 miles away.

1898 (June 3)—First paid radio message is sent from the Needles, Isle of Wight station.

1898 (July 20)—Kingstown regatta off Ireland is reported by wireless to a Dublin newspaper from the steamer *Flying Huntress*.

1899—Elster and Geitel discover that various elements possess photoelectric properties.

1899 (March 27)—Marconi signals by wireless across the English Channel for the first time.

1899—Marconi proves curvature of the earth does not interfere with propagation of wireless waves.

1899 (April 22)—The first French gunboat is equipped with wireless at Boulogne.

1899 (April 28)—Steamer *R. F. Mathews* collides with East Goodwin Sands Lightship and flashes the first wireless call for assistance.

1899 (April)—United States Army Signal Corps establishes wireless communication between Fire Island and Fire Island Lightship, a distance of twelve miles, and later between Governor's Island and Fort Hamilton.

1899 (July)—British warships *Alexandra, Juno* and *Europa* exchange wireless messages at sea up to seventy-five nautical miles.

1900—Sir Oliver Heaviside (died February 4, 1925) and Professor Arthur E. Kennelly suggest theory of "radio mirror" now known as the Heaviside surface, a conducting medium in the upper levels of the atmosphere.

1900—A. F. Collins uses his so-called "electrostatic system" to signal eight miles by wireless.

1900 (February 18)—First German commercial wireless station is opened on Borkum Island.

1900 (February 28)—S. S. *Kaiser Wilhelm der Grosse* equipped with wireless and leaves port as the first seagoing passenger vessel to carry such service. Borkum Island hears it sixty miles away.

1900—Michael Pupin invents the loading coil that improves long distance telephony.

1900—Marconi files application for his famed Patent 7777 for a "tuned" or synchronized system of wireless.

1900 (November 2)—Belgium's first wireless station is completed at Lapanne.

1901 (January 1)—The bark *Medora* is reported by wireless to be waterlogged on Ratel Bank and assistance is sent.

APPENDIX 271

1901 (February 11)—Wireless communication across 196 miles is established between Niton station, Isle of Wight, and the Lizard station.

1901 (March)—Public wireless service inaugurated between five principal islands of the Hawaiian group.

1901 (September 28)—Professor Reginald A. Fessenden applies for United States patent on "improvements in apparatus for wireless transmission of electromagnetic waves, said improvements relating more especially to the transmission and reproduction of words or other audible signals." He contemplates use of an alternating current generator having a frequency of 50,000 cycles a second.

1901—Dr. John Stone applies for United States patents covering wireless telegraphy.

1901 (December 12)—Marconi, with two assistants, P. W. Paget and G. S. Kemp, at St. Johns, Newfoundland, picks up the first transatlantic wireless signal, the letter "S" sent from the transmitter at Poldhu.

1902 (February)—Marconi on S. S. *Philadelphia* hears signals from Poldhu 2,099 miles away.

1902 (June 25)—Marconi introduces the magnetic detector, actuated by clockwork on the Italian cruiser *Carlo Alberto*.

1902—Electrolytic detector introduced by Professor R. A. Fessenden.

1902—Professor E. Ruhmer's photophone system of wireless covers a distance of twenty miles at Kiel, Germany.

1902 (July 14)—Marconi on Italian cruiser *Carlo Alberto*, at Cape Skagen, receives a message from Poldhu 800 miles distant and from Kronstadt, 1600 miles.

1902 (December 17)—First west-east transatlantic wireless messages sent by Marconi from Glace Bay to England.

1903—Valdemar Poulsen and William Duddell introduce the electric arc transmitter as a means of propagating electromagnetic waves.

1903—Message from President Roosevelt to King Edward of England sent via station WCC, South Wellsfleet, Cape Cod, is received at Poldhu.

1903—First ocean daily newspaper instituted on board S. S. *Campania*, with dispatches supplied by wireless.

1903 (August 4)—First International Radiotelegraphic Conference held at Berlin.

1903—Poulsen patents an improved arc oscillation generator using a hydrocarbon atmosphere and a magnetic field.

1904 (February 1)—Marconi Company institutes CQD as the wireless call of distress.

1904—Professor John Ambrose Fleming, of England, invents the two-element thermionic valve detector, the patent number being 24850.

1904 (August 15)—Great Britain passes a wireless telegraph act.

1904—Wireless apparatus displayed as one of the marvels at St. Louis World's Fair.

1905—Marconi patents a horizontal directional transmitting aerial and predicts that he will soon be able to reach the antipodes more easily than nearby places.

1905—The *New York Times* receives eyewitness wireless reports of naval battle off Port Arthur in Russo-Japanese war.

1906—E. Bellini and A. Tosi in Italy pioneer in radio direction-finder research.

1906—Rignoux and Fournier, French physicists, use selenium cells to construct artificial retina. Each cell is linked by wire to a shutter that opens when light actuates the cell.

1906—Telefunken arc system of wireless telegraphy is developed and covers a distance of twenty-five miles.

1906—Lee de Forest invents the three-element vacuum tube that has a filament, plate and grid.

1906—Professor R. A. Fessenden develops a high frequency alternator and installs it at Brant Rock, Mass., for communication with ships at sea.

1906—Dunwoody discovers the rectifying properties of carborundum crystals and Greenleaf Pickard discovers similar properties of silicon.

1907—Coherer replaced by the crystal, magnetic, thermal and electrolytic detectors.

1907 (January 18)—Lee de Forest is granted a patent on the three-element vacuum tube which he calls "the audion."

1907—Arthur Korn sends a picture of President Fallières of France by wire from Berlin to Paris in twelve minutes.

1907 (October 17)—Commercial wireless service begins between Clifden, Ireland, and Glace Bay, Nova Scotia.

1908 (February 2)—S. S. *St. Cuthbert* on fire off Sable Island is sighted by S. S. *Cymric* from which a newspaper correspondent sends story by wireless to *The New York Times* and *Chicago Tribune*.

1908 (February 3)—Marconi transatlantic wireless stations opened to the public for transmission and reception of Marconigrams between England and Canada.

APPENDIX

1908—Professor Marjorana designs an arc oscillating generator and liquid microphone system utilizing it for communication between Rome and Sicily.

1908—Fessenden constructs a high frequency alternator with an output of 2.5 kilowatts at 225 volts and with a frequency of 70,000 cycles a second.

1908—Telefunken Company conducts a series of tests between Sandy Hook and Bedloe's Island to prove practicability of the radiophone.

1908—International Radio Telegraphic Conference at Berlin proposes SOS as wireless distress call instead of CQD.

1908—Poulsen develops an arc transmitter that covers 150 miles on the first test.

1909 (January 23)—S. S. *Republic* collides with S. S. *Florida* off New York; Jack Binns, the wireless operator of the *Republic*, sends the CQD and summons assistance thereby proving the value of radio in time of disaster at sea.

1909—Marconi awarded Nobel Prize in physics.

1910 (January 7)—Steamship *Puritan* caught in ice in Lake Michigan flashes SOS and tugs go to the rescue of fifteen passengers.

1910 (January 13)—Enrico Caruso and Emmy Destinn sing in deForest radiophone broadcast from Metropolitan Opera House. It is picked up by S. S. *Avon* at sea and in Bridgeport, Conn.

1910—A. Ekstrom of Sweden discovers that he can "scan" an object directly by use of a strong beam of light behind a scanning disk.

1910—Marconi sends wireless to Buenos Aires from Ireland.

1910—S. S. *Principessa Mafalda* receives Clifden signal across a distance of 4,000 miles by day and 6,735 at night.

1910 (April 23)—Marconi transatlantic America-Europe service opened.

1910 (June 24)—United States government approves act requiring radio equipment and operators on certain passenger carrying vessels.

1911 (July 1)—Department of Commerce organizes radio division to enforce act of June 24, 1910.

1911—Radio telephone covers 350 miles between Nauen, Germany, and Vienna, Austria.

1912—Frederick A. Kolster, of the Bureau of Standards, develops a decremeter to make direct measurements of radio wave lengths.

1912—United Wireless Company is absorbed by the American Marconi Co.

1912 (February)—Marconi Company procures Bellini Tosi patents including the direction finder.

1912 (February 3)—First Australian Commonwealth wireless station is opened.

1912 (April 14)—S. S. *Titanic* disaster proves the value of wireless at sea. Seven hundred lives are saved.

1912 (July 5)—International Radio Telegraphic Conference in London approves regulations to secure uniformity of practice in radio services.

1912—United States Naval Radio Station NAA opens at Arlington, Va.

1912—Edwin H. Armstrong develops a regenerative vacuum tube circuit while experimenting at Hartley Laboratory, Columbia University.

1912—Marconi patents "the timed spark system" by which exceedingly long waves can be employed (14,000 meters and longer).

1912 (July 23)—Act approved by United States government extending act of June 24, 1910, to cover cargo vessels and requiring auxiliary source of power, efficient communication between wireless cabin and bridge, and two or more skilled wireless operators in charge of apparatus on certain passenger ships.

1912 (August 13)—United States government approves act licensing radio operators and transmitting stations.

1913—United States and French governments cooperate between Arlington and Eiffel Tower to procure data for comparing velocity of electromagnetic waves with that of light.

1913 (June)—Radiotelegraph Act of Canada passed by Parliament at Ottawa.

1913—Station POZ, Nauen, Germany, sends a message 1,550 miles.

1913—Dr. William David Coolidge invents "hot" cathode ray tube and makes useful developments in X-ray tubes.

1913 (September)—Prince Albert, ruler of principality of Monaco, sails into New York harbor on his yacht *Hirondelle* equipped with a wireless piano.

1913 (October 11)—S. S. *Volturno* on fire at sea. Wireless call for help brings ten vessels to the rescue.

1913—Wireless station at Macquerie Island keeps Dr. Mauson, Australian explorer, in communication with outer world.

APPENDIX

1913 (November 12)—Safety at Sea Conference held in London and wireless receives major consideration.

1913 (November 24)—Wireless tests made on Delaware, Lackawanna & Western Railroad between Hoboken and Buffalo.

1914—Direct communication established between WSL, Sayville on Long Island and POZ, Nauen, Germany, and between Tuckerton-Elvise.

1914—Two warships at sea report radio telephone reliable for communication over 18 miles.

1914 (April 15)—Memorial unveiled at Godalming in honor of Jack Phillips, chief operator of ill-fated *Titanic* who died at his post.

1914—Motor lifeboats of S. S. *Aquitania* are equipped with wireless marking a new departure in the application of radio to safety of life at sea.

1914 (September 24)—California-Honolulu wireless circuit opened by the Marconi Wireless Telegraph Company of America.

1914—Laws formulated by foremost maritime nations requiring vessels of certain size and grades to carry wireless apparatus and operators.

1914—United States District Court, Eastern District of New York, in opinion handed down by Judge Van Vechten Veeder upholds validity and priority of Marconi's patents.

1914—Cryptic wireless message from Nauen, Germany, tells *Kronprinzessin Cecile* 850 miles off Irish coast to dash for a neutral port with the $10,000,000 gold on board. It surprises Bar Harbor by arriving there several days later.

1914 (October 6)—E. H. Armstrong issued a patent covering the regenerative circuit known as the feed-back or self-heterodyne circuit.

1914—Marconi turns his attention to adapting radio to warfare including short waves, secret communication, direction finders, and "narrow-casting" by the use of parabolic reflectors and radio beams.

1915 (February 20)—Panama-Pacific Exhibition at San Francisco is officially opened by President Wilson at Washington, through wireless signal.

1915 (May 12)—Monument in Battery Park, New York, unveiled in honor of wireless operators who lost their lives at post of duty.

1915 (May 22)—Marconi predicts visible telephony as he sails from New York for Rome upon request of King Victor Emmanuel because of Italy's entry into World War.

1915—Dr. F. A. Kolster at the Bureau of Standards develops a moveable coil type radio compass.

1915 (July 27)—Wireless communication established between United States and Japan via relay through Honolulu.

1915 (July 28)—American Telephone and Telegraph Company working in conjunction with Western Electric engineers at Arlington, Va., succeeds in telephoning by radio to Paris, 3,700 miles, and to Hawaii, 5,000 miles.

1916—Determination of the difference in longitude between Paris and Washington with assistance of radio which has been in progress since 1913 is completed. The result, expressed in terms of time being 5 hours 17 minutes and 35.67 seconds, has a probable accuracy of 0.01 second.

1916 (November 5)—President Wilson and Mikado of Japan exchange radiograms at opening of transpacific circuit.

1916 (November)—DeForest experimental radiophone station opens at High Bridge, N. Y.

1916 (November)—Station 2ZK, New Rochelle, operated by George C. Cannon and Charles V. Logwood, broadcasts music between 9 and 10 P.M., daily except Sunday.

1917—E. F. W. Alexanderson designs 200-kilowatt high frequency alternator making world-wide wireless possible.

1917—German submarines elude Allied listening posts by use of short waves (75 meters).

1917 (June 2)—Wireless "becomes of age" in England. Twenty-one years have passed since the registration of wireless patent No. 12039 in 1896.

1918—A. Hoxie installs high-speed wireless recorder at Otter Cliffs, Me., to copy messages from France.

1918—Radiotelegraph and radiophone conclusively prove their tremendous importance in warfare during the World War.

1918—Progress toward continuous-wave radio as distinct from damped waves is marked, chiefly because of the vacuum tube as a generator of undamped oscillations. Wireless telephony also forges ahead.

1918—High power radio station built by the United States is opened at Croix d'Hins, near Bordeaux. It is called the Lafayette station.

1918—Erection of a high power station near Buenos Aires is begun. It will communicate direct with North America.

1918 (April)—High power station LCM, opened at Stavanger, Norway, for use of Norwegian government. The signal is clear in the United States.

APPENDIX

1918—Application of wireless to ships continues and at the end of the year between 2,500 to 3,000 vessels in the British Merchant Marine carry transmitters and receivers.

1918 (July 31)—United States government takes over all wireless land stations in the country, with exception of a few high power transmitters which remain under control of commercial organizations.

1918 (September 22)—Sydney, Australia, hears wireless from Carnarvon, England, 12,000 miles. Confirmation of the dispatches sent by cable at the same time arrive several hours later.

1918 (November)—Wireless from France and Germany announces signing of the Armistice.

1919 (February)—Spanish decree specifies that all sailing vessels of 500 tons or more and carrying fifty or more passengers must be wireless equipped.

1919—The "spark" and "arc" era in radio transmission begin to give way to the vacuum tube.

1919—President Wilson goes to Peace Conference in Paris while wireless on board the S. S. *George Washington* maintains communication with shore.

1919—NC-flying boats use radio on transatlantic flight.

1919 (June 30)—There are 2,312 ship stations licensed by the United States, an increase from 1,478 since June 30, 1918, chiefly due to number of vessels built for war.

1919 (August 24)—United States Signal Corps broadcasts service of Trinity Church at Third and D streets, Washington, D. C.

1919—British Parliament passes bill specifying that all merchant vessels of 1,600 tons or more under English flag must carry wireless. This makes permanent a temporary war measure.

1919—British dirigible R-34 crosses Atlantic equipped with a vacuum tube transmitter.

1919—Radiophone links England and Canada by use of vacuum tube transmitters.

1919—President Wilson returning from Peace Conference on board S. S. *George Washington* makes Memorial Day address to crew and his voice is heard in a broadcast to shore.

1919—Radio Corporation of America organized, taking over the interests of the Marconi Wireless Telegraph Company of America and radio activities of the General Electric Company in plans for an American world-wide radio system.

1920 (January 14)—Greece passes a law that makes carrying wireless equipment obligatory on all Greek merchant ships of

1,600 gross tons or over, or having fifty persons on board including the crew.

1920 (January 25)—High power station LPZ opened at Mont-Grande, Argentina.

1920 (February 29)—United States government returns high power stations under its control during World War, and first commercial long distance radio communication between the United States and foreign countries is inaugurated by the Radio Corporation of America.

1920—A tract of land covering ten square miles is acquired on Long Island at Rocky Point and Riverhead for the construction of a Radio Central conceived for world-wide communication.

1920—American radio amateurs reorganize their forces, now reinforced many thousands of times by war-trained radio men, and begin to turn their attention to amateur radiophone.

1920—Installation of 200-kilowatt Alexanderson high frequency alternators for international communication begins at Bolinas, Calif., Marion, Mass., and Kahuku, Hawaii.

1920 (November 2)—Radio broadcasting begins with KDKA, Pittsburgh, the pioneer station broadcasting Harding-Cox election returns.

1921—President Harding formally opens Radio Central on Long Island by sending a radiogram addressed to all nations.

1921—Paul Godley goes to Ardrossan, Scotland, and hears twenty-seven radio amateurs in the United States make history in their field by transmitting across the Atlantic on power outputs ranging from 50 to 1,000 watts.

1921—200-kilowatt Alexanderson alternator system installed at Tuckerton, N. J.

1921 (July 2)—Dempsey-Carpentier fight is broadcast from Boyle's Thirty Acres in Jersey City, N. J., by a temporarily installed transmitter at Hoboken.

1921—Professor Edouard Branly awarded Nobel Prize for Physics because of his radio research work.

1921 (August 30)—First annual convention of American Radio Relay League held in Chicago.

1921 (September 27)—Station WBZ opens at Springfield, Mass.

1921 (October 1)—Station WJZ officially opened at Newark, N. J., as the first broadcaster in the metropolitan area. First program features World Series bulletins.

1921 (November 11)—Burial of unknown soldier at Arlington including an address by President Harding is broadcast.

1921 (November 11)—Station KYW goes on the air in Chicago.

APPENDIX 279

1921 (December 15)—Broadcasting station WDY opens at Roselle Park, N. J., (continued until February 15, 1922, when it was amalgamated with WJZ previously opened at Newark).

1922—First ship-to-shore two-way radio conversation between Deal Beach, N. J., and S. S. *America* 400 miles at sea.

1922—S. S. *Gloucester* off Jersey coast talks to Deal Beach, N. J., which relays voices by wire to Long Beach, Calif., and then by radiophone to the Catalina Islands.

1922 (February 20)—Station WGY, Schenectady, goes on the air.

1922 (February 27)—First Annual Radio Conference, pertaining to broadcasting, held in Washington, D. C.

1922—Marconi demonstrates to Institute of Radio Engineers his radio beam system of communication that utilizes reflectors to concentrate radio energy in much the same way that a searchlight casts a beam of light.

1922 (July 25)—Station WBAY abandoned by the American Telephone and Telegraph Co.

1922 (August 16)—Station WEAF goes on the air with transmitter atop Western Electric Building on West Street, New York.

1922—Edwin H. Armstrong announces his superheterodyne and super-regenerative circuits.

1922 (September 7)—First commercial broadcast over WEAF sponsored by the Queensborough Corporation.

1922 (October 15)—First time in history high-powered vacuum tube transmitters handle traffic between New York, England and Germany.

1922 (October 28)—Princeton-Chicago football game goes on the air as the first gridiron broadcast.

1922 (November 11)—Remote control pick-up of opera *Aïda* from Kingsbridge Armory, New York.

1922 (November 22)—First broadcast by New York Philharmonic-Symphony Orchestra.

1922—Dr. Irving Langmuir of the General Electric Company announces a 20-kilowatt vacuum tube.

1923 (January 4)—First chain broadcast with telephone lines connecting WEAF, New York, with WNAC, Boston.

1923 (March)—Professor L. A. Hazeltine describes his invention of the neutrodyne circuit at Radio Club of America meeting.

1923—C. Francis Jenkins sends a picture of President Harding by television from Washington to Philadelphia.

1923 (March 4)—Station KDPM, Cleveland, Ohio, picks up short waves from KDKA, Pittsburgh, and thereby stages the first rebroadcast program.

1923 (March 20)—Second Annual Radio Conference held in Washington, D. C.

1923—Radio station built in a valley between the Herzogstand and the Stein, two foothills of the Bavarian Alps, features an aerial suspended by wire cables stretched between the tops of the two peaks.

1923—Increased radio traffic to and from ocean liners leads to installation of high speed transmitters and automatic reception.

1923 (May 15)—Station WJZ moves from Newark to New York.

1923 (June)—First multiple station network with WEAF, New York, WGY, Schenectady, KDKA, Pittsburgh, and KYW, Chicago, linked by wires.

1923—President Warren G. Harding speaks from St. Louis as he begins the western tour that ends in his death at San Francisco. The stations are WJZ, New York; WCAP, Washington; KSD, St. Louis.

1923 (August 1)—Station WRC opened at Washington, D. C.

1923—American and French amateurs establish two-way communication across Atlantic on 100-meter wave.

1923—Charles Proteus Steinmetz declares "there are no ether waves."

1923 (November 11)—Woodrow Wilson's Armistice Day address broadcast by WEAF, his only public address after retiring from the White House.

1923—Wireless controlled airplane makes flight without a pilot at the Etampes Aerodrome in France. Flights were also made with a pilot using a gyroscopic stabilizer and special steering motors controlled from the ground.

1923—International Commission for Aerial Navigation agree, as a general principle, all aircraft engaged in public transport should carry radio equipment.

1923—Tube delivering 20 kilowatts of high frequency energy to the aerial is introduced.

1923 (December 4)—First broadcast from United States Capitol, opening of Congress.

1923—Donald B. MacMillan in Arctic region uses short waves from ship, the *Bowdoin*, to communicate with Chicago, New York and other cities. He hears broadcasting stations in United States and England.

1924 (January 9)—Station KGO, Oakland, Calif., goes on the air.

1924—The 800-kilowatt station at Monte Grande, Argentina, is opened for communication with New York, Paris and Berlin.

APPENDIX 281

1924 (February 5)—England rebroadcasts a short-wave program sent across the sea by KDKA.

1924 (February 6)—Funeral services for Woodrow Wilson at National Cathedral, Washington, D. C., are broadcast with WEAF as the New York outlet.

1924 (February 23)—Calcutta picks up KDKA program relayed from London.

1924 (May 30)—Marconi using short waves talks by voice from his yacht off England to Australia.

1924—National Republican convention at Cleveland and National Democratic convention at New York broadcast by nation-wide networks.

1924 (July)—British government and Marconi Wireless Telegraph Co., plan to link the Empire by beam radio system.

1924—Marconi in lecture before the Royal Society of Arts describes his short-wave beam system.

1924 (September)—Marconi using the 32-meter wave in daylight talks with Syria by voice from his yacht 2,100 miles away.

1924 (October)—Zeppelin ZR-3 (renamed *Los Angeles*) crosses Atlantic equipped with wireless.

1924—Wireless "lighthouse" established on an island in the Firth of Forth, Scotland. The wireless energy concentrated by reflectors flashes a beam that ships within a 100-mile area can detect to determine their position in fog.

1924 (October 6)—Third National Radio Conference held in Washington, D. C.

1924 (October 11)—Cape Town, Africa, intercepts program from KDKA and rebroadcasts it.

1924 (November 30)—Pictures of President Coolidge, Prince of Wales, Premier Stanley Baldwin and others sent by facsimile radio from London to New York in twenty minutes.

1924—First international broadcast with program transmitted on long wave (1,600 meters) from Coventry, England, picked up at Houlton, Me., transmitted by wire to WJZ and rebroadcast.

1924 (December 15)—Station KOA, Denver, goes on the air.

1925—Commercial applications of short waves progress as transatlantic traffic is handled on channels from 20 to 105 meters.

1925—Trend toward high power broadcasting sends the transmitters outside the thickly populated areas to minimize interference.

1925—Experiments are conducted at Pittsburgh and Schenectady with 50-kilowatt transmitters for broadcasting.

1925—Three-meter waves generated at Technical Physical Institute at Jena with a capacity of about 100 watts.

1925—Coolidge inaugural broadcast by twenty-four stations scattered from coast-to-coast.

1925 (April)—Radio shadowgraphs demonstrated by John L. Baird in Selfridge store, London.

1925 (May 7)—Facsimile messages, maps and pictures radioed from New York to Honolulu, 5,136 miles, by the Ranger photoradio system.

1925—Nichols and Schelling of Bell Telephone Laboratories suggest theory to account for fading of radio, which they believe is caused by the earth's magnetic field's effect on wave propagation.

1925—Stations WJZ and WRC rebroadcast the sound of Big Ben atop the House of Parliament when it strikes midnight.

1925—Radio receiving sets and tubes designed for complete alternating current operation are introduced for home use.

1925—United States Naval Radio Research Laboratory at Bellevue, D. C., and Carnegie Institution confirm Heaviside-Kennelly theory.

1926 (January 1)—John McCormack and Lucrezia Bori make their radio débuts over WJZ, a move that encourages other noted artists to go on the air.

1926—S. S. *President Roosevelt* successfully uses radio compass in blinding snowstorm to find S. S. *Antinoe* in distress.

1926 (February 23)—President Coolidge signs the Dill-White Radio Bill.

1926 (April 20)—Picturegram of check sent from London to New York where it is honored and cashed.

1926 (May)—Byrd and Bennett in plane *Josepine Ford* fly to the North Pole from Spitzbergen carrying a 44-meter radio transmitter to maintain contact with the base.

1926 (May)—Dirigible *Norge* sails over the Arctic and sends wireless message direct from the North Pole.

1926—Radio receiving sets having complete alternating current or light socket operation are introduced for home use.

1926 (September 23)—Dempsey-Tunney fight is broadcast by long and short waves to all parts of the world.

1926—World Series is broadcast by WJZ's national network.

1926 (November 1)—National Broadcasting Company organized.

1926 (December 15)—Alexanderson in St. Louis demonstrates an advance in television by showing his multiple light-brush system and new projector.

APPENDIX

1927 (January 1)—Initial coast-to-coast hook-up using a 4,000-mile network to broadcast football game in Bowl of Roses, California.

1927 (January 7)—Radiotelephone circuit opens between New York and London. Adolph S. Ochs, publisher of the *New York Times,* talks with Geoffrey Dawson, editor of the London *Times.*

1927 (January 21)—First coast-to-coast broadcast of opera (*Faust*) from stage of the Chicago Civic Auditorium.

1927 (February 3)—John L. Baird describes his television system at Glasgow.

1927 (February 22)—First coast-to-coast Presidential broadcast and first from the floor of Congress; Washington Birthday address by Calvin Coolidge at joint session of Congress.

1927 (March 2)—Federal Radio Commission is appointed; Rear Admiral W. H. G. Bullard, John F. Dillon, Judge E. O. Sykes, O. H. Caldwell and H. A. Bellows.

1927 (April 7)—Wire television demonstrated between Washington, D. C., and New York; and radio television between Whippany, N. J., and New York by Bell Telephone Laboratories.

1927 (June 11)—Massachusetts Institute of Technology dinner in New York sees photoradio messages and pictures arrive from London and Hawaii.

1927—Arrival of Lindbergh back in United States after historic flight to Paris is broadcast by largest network of stations ever assembled up to this time.

1927—Plane *America* with Byrd, Balchen, Acosta and Noville hops off for Europe with radio equipment on board.

1927 (August 20)—Airplane *Dallas Spirit* in tail spin over Pacific on way to Hawaii flashes SOS on 33-meter wave which is picked up by the *New York Times'* receiving station, 3,500 miles away.

1927 (September 18)—Columbia Broadcasting System goes on the air with a basic network of sixteen stations.

1927 (October 17)—Marconi predicts at Institute of Radio Engineers that short waves are destined to play a vital rôle in radio progress and television.

1927 (December 30)—Radiomarine Corporation of America organized to operate radio service for ships at sea.

1928 (February 8)—Mrs. Mia Howe in London is televised by Baird and is seen in Hartsdale, N. Y., as the first television face to cross the Atlantic.

1928 (March 7)—Passengers on S. S. *Berengaria* 1,000 miles distant see face of Dora Selvy televised in London.

1928 (July 12)—Televising of outdoor scenes without use of artificial light is accomplished at the Bell Telephone Laboratories.

1928 (August 11)—Hoover is officially notified of his nomination for presidency while 107 stations are linked with the microphones at Palo Alto, Calif.

1928 (September 11)—A one-act melodrama, *The Queen's Messenger*, is televised at Schenectady.

1929 (February 1)—Band concert from Queens Hall, London, broadcast as the first scheduled international rebroadcast.

1929 (February)—While D. W. Griffith broadcasts at Schenectady he is televised and seen in Los Angeles by radio.

1929 (June)—Thanksgiving service at Westminster Abbey for recovery of King George is rebroadcast in United States.

1929—Screen-grid tube permitting greater sensitivity of receiving set with fewer tubes is developed.

1929 (June 27)—Television in color demonstrated by Bell Telephone Laboratories by wire from one end of a room to the other.

1929 (August 15)—Brokerage offices established on several ocean liners are supplied Wall Street service by wireless.

1929 (November 15)—Radio handles efficiently and expeditiously greatly increased volume of transatlantic communications when earthquake snaps twelve cables on bed of North Atlantic.

1929 (November 18)—Zworykin demonstrates his kinescope or cathode ray television system at Rochester, N. Y.

1929 (November 29)—Short wave radio from Little America, Antarctica, announces that Byrd flew over the South Pole. Balchen piloted the machine.

1929 (December 20)—First international program from Germany, broadcast from Koenigswusterhausen by short wave and rebroadcast by stations in United States.

1929 (December 25)—International exchange of programs between United States, Germany, England and Holland.

1929—Dr. Karolus of Germany contributes an electro-chemical light valve or "shutter" to television so more powerful illumination can be used.

1930 (January 21)—King George V welcomes delegates to the London Naval Conference and is heard in his first worldwide broadcast.

APPENDIX

1930 (February 18)—Drawing of rectangular design is sent by television to Australia and flashed back to Schenectady without losing its identity.

1930 (March 11)—Arrival of Byrd Antarctic Expedition at Dunedin, New Zealand, and two-way conversation between members of the expedition and friends in New York heard in a rebroadcast throughout the United States.

1930 (April 6)—John L. Baird televises "abbreviated vaudeville" in London.

1930 (April 9)—Two-way wire television in which speakers at ends of 3-mile line see each other as they converse is demonstrated by Bell Telephone Laboratories.

1930 (April)—U. A. Sanabria shows television images on a two-foot screen in his Chicago laboratory.

1930 (April 30)—Two-way radiophone conversation between Marconi aboard his yacht near Italian coast and friends in New York.

1930—The pentode and supercontrol tubes for broadcast reception are introduced.

1930 (May 22)—Television is seen on six-foot screen in Proctor's theater in Schenectady.

1930 (June)—S.S. *America* off Fastnet Island, approximately 3,000 miles from New York, picks up facsimile messages from United States.

1930 (June)—Plans announced for $250,000,000 Radio City to be built on Manhattan Island.

1930 (June 10)—John Hays Hammond, Jr., describes his patent for a television eye for airplanes that enables pilots to "see" through fog and darkness to make safe landings.

1930 (June 30)—First round-the-world broadcast, Schenectady to Holland, relay to Java, Australia and back to point of origin in less than a second.

1930 (July 20)—Play, *The Man with a Flower in his Mouth*, televised in London while dramatic critics watch.

1930 (July 30)—Religious program in Nidaros Cathedral, Norway, in celebration of 900th anniversary of introduction of Christianity in Norway rebroadcast in United States.

1930 (September 14)—Provisional President Uriburu of the Argentine Republic addresses American people by radio from Buenos Aires.

1930 (December 6)—Direct radio communication established with China by opening of circuit between San Francisco and Shanghai.

1930 (December 14)—Farnsworth informs Federal Radio Commission he has succeeded in narrowing wave band required for television to 6,000 cycles width.

1930 (December 25)—Japan is heard in first American rebroadcast from the Orient with Premier Hamaguchi as the speaker.

1931 (January 1)—Voice of Benito Mussolini, Italian Premier, is heard in the United States for the first time in an international broadcast over short wave station in Rome.

1931 (January 11)—Cæsium photoelectric cells that "see red" introduced by Bell Telephone Laboratories to clarify the images.

1931 (February 12)—Pope Pius XI addresses the world in an international broadcast inaugurating Vatican City station HVJ. First time Pope's voice is heard in America.

1931 (March 31)—Micro-rays (18 cm.) carry voices across the English Channel between Dover and Calais.

1931 (April 26)—Television station W2XCR goes on the air in New York.

1931 (April 29)—Representatives of new Spanish Republic broadcast greetings to the United States from Madrid.

1931 (May 15)—Program originating in Bangkok, Siam, sent by short wave to United States and rebroadcast for pleasure of Siam's King visiting in New York.

1931 (May 25)—Argentine Independence Day celebration is rebroadcast in United States.

1931 (June 3)—English Derby televised for the first time by John L. Baird at Epsom Downs.

1931 (June)—Empire State Building, world's highest skyscraper, is selected as the site for a television station that will use quasi-optical waves.

1931 (July 21)—Experimental television station W2XAB opened in New York.

1931 (August 21)—Vienna Philharmonic Orchestra is rebroadcast in America by WJZ.

1931 (September 13)—Mahatma Gandhi, "India's man of destiny," explains the political and economic plight of his country to America in a rebroadcast from London.

1931 (September 24)—Sanabria demonstrates television on 10-foot screen at Radio-Electrical World's Fair in New York.

1931 (October)—Professor Jacob Papish and Eugene Wainer of Cornell University discover element No. 87 in mineral

APPENDIX 287

samarskite. It is said to be similar to cæsium and may greatly increase sensitivity of photoelectric cells.

1931 (October 22)—Television on 10-foot screen is shown at the Broadway Theatre, New York, with 1,700 attending the opening performance. A wire link is used to the televisor in the Theatre Guild Playhouse.

1931 (October 27)—Marconi experimenting on the Ligurian coast near Genoa with 50 centimeter waves.

1931 (November)—Television images from Chicago are picked up at unemployment relief bazaar at Ottumwa, Iowa, 250 miles away.

1931 (November)—Alexanderson sends television across his laboratory on a beam of light instead of a radio wave or wire.

1931 (December 12)—Thirtieth anniversary of first transatlantic wireless signal is celebrated by a world-wide broadcast featuring tributes to Marconi from fifteen nations and insular possessions.

1931 (December 25)—*Hansel und Gretel* is broadcast from the Metropolitan Opera House as the first radio presentation by that organization. Combined networks of WEAF and WJZ are linked with the microphones.

1932 (February)—Delegates and radio observers at World Disarmament conference at Geneva are heard in rebroadcasts from Switzerland.

1932 (February 22)—World-wide tributes to Washington on Bicentennial of his birth are heard in America including address by President Paul Doumer of France at American Club in Paris.

1932 (March)—Radio broadcasting facilities mobilized to aid search for kidnapped Charles A. Lindbergh, Jr., and to flash bulletins to the anxious public.

1932 (March)—Jenkins describes new television principle. Images said to be 3,600 times brighter than heretofore, appear on a sensitized emulsion of "an animated lantern slide." Incoming signals quickly change the surface from opaque to clear, equivalent to light and shade, thereby "painting" an ever-changing pattern, corresponding to the scene at the transmitter.

1932 (March 13)—German Presidential election returns, Paul Von Hindenburg versus Adolph Hitler are rebroadcast in United States.

1932 (April 7)—Marconi announces successful tests with ultra-short waves and reports that he expects soon to be able to see his family in New York while he speaks with them by radiophone.

TELEVISION STATIONS IN UNITED STATES

Call Letters	Power (Watts)	Company	Location
		2000–2100 kc.	
W3XK	5000	Jenkins Laboratories	Silver Spring, Md.
W2XCR	5000	Jenkins Television Corp.	New York, N. Y.
W2XAP	250	" " "	Portable
W2XCD	5000	DeForest Radio Company	Passaic, N. J.
W9XAO	500	Western Television Corp.	Chicago, Ill.
W9XAA	1000	Chicago Federation of Labor	Chicago, Ill.
		2100–2200 kc.	
W3XAK	5000	National Broadcasting Co.	Portable
W2XBS	5000	" " "	New York, N. Y.
W3XAD	500	RCA Victor Company	Camden, N. J.
W2XCW	20,000	General Electric Co.	S. Schenectady, N. Y.
W8XAV	20,000	Westinghouse E. & M. Co.	E. Pittsburgh, Pa.
W9XAP	2500	Chicago Daily News	Chicago, Ill.
W2XR	500	Radio Pictures, Inc.	Long Island City, N. Y,
W6XS	500	Don Lee, Inc.	Garden City, Calif.
		2750–2850 kc.	
W2XBO	500	United Research Corp.	Long Island City, N. Y.
W9XG	1500	Purdue University	W. Lafayette, Ind.
W2XAB	500	Atlantic Broadcasting Corp.	New York, N. Y.
		2850–2950 kc.	
W1XAV	1000	Shortwave & Television Laboratories, Inc.	Boston, Mass.
W2XR	500	Radio Pictures, Inc.	Long Island City, N. Y
		43,000–46,000 kc.	
W9XD	500	The Journal Company	Milwaukee, Wis.
W3XAD	2000	RCA Victor Company, Inc.	Camden, N. J.
W2XBT	750	National Broadcasting Co.	Portable
W1XG	30	Shortwave & Television Co.	Portable
W2XR	1000	Radio Pictures	Long Island City, N. Y.
W2XF	5000	National Broadcasting Co.	New York, N. Y.
W6XAO	1000	Don Lee, Inc.	Los Angeles, Calif.

APPENDIX

48,500–50,300 kc.

W9XD	500	The Journal Company	Milwaukee, Wis.
W3XAD	2000	RCA Victor Company, Inc.	Camden, N. J.
W2XBT	750	National Broadcasting Co.	Portable
W1XG	30	Shortwave & Television Co.	Portable
W2XR	1000	Radio Pictures	Long Island City, N. Y.
W2XF	5000	National Broadcasting Co.	New York, N. Y.

60,000–80,000 kc.

W9XD	500	The Journal Company	Milwaukee, Wis.
W3XAD	2000	RCA Victor Company, Inc.	Camden, N. J.
W2XBT	750	National Broadcasting Co.	Portable
W1XG	30	Shortwave & Television Co.	Portable
W2XR	1000	Radio Pictures	Long Island City, N. Y.
W2XF	5000	National Broadcasting Co.	New York, N. Y.

CANADIAN TELEVISION STATIONS

Call	Owner	City	Kilocycles
VE9RM	Rogers Majestic Corp., Ltd.	Toronto, Ont.	2004–2100
VE9EC	LaPresse Publishing Co.	Montreal, Que.	2004–2100
VE9DS	Canadian Marconi Company	Mount Royal, Que.	2100–2200
VE9BZ	Radio Service Engineers	Vancouver, B. C.	2750–2850
VE9AR	A. R. MacKenzie	Saskatoon, Sask.	2850–2950
VE9AF	J. A. Ogilvy's Limited	Montreal, Que.	2850–2950
VE9ED	Dr. J. L. P. Landry	Mont Joli, Que.	2850–2950

INDEX

Advertising, by television, 227-229
Aircraft, applications of television to, 125, 244-245
   Hammond's television "eye" for, 127-134
Alexanderson, Dr. E. F. W., describes his television system, 58-62
   future of television as seen by, 62-64, 88-89, 122-126
   image sent to Australia by, 106-108
   images sent to Germany by, 202
   images transmitted on light beam by, 212-214
   one-act play televised by, 87-90
   television demonstrated on theater screen by, 118-122
   television research directed by, 12
Amstutz, N. S., picture transmission by, 7
Ardenne, Manfred von, television experimenter, 55
Argon tubes, use in television, 96-97
Aylesworth, M. H., television possibilities discussed by, 140-141, 199-201, 240-241

Bain, Alexander, picture transmission experimenter, 5-6
Baird, Hollis, contribution to television, 13
   electrical and mechanical scanning compared by, 104-105
   ultra-short waves discussed by, 190
   uses of television foreseen by, 196-197
Baird, John L., career of, 82-84
   contributions to television, 12
   defines television, 73
   describes his system, 55-57
   English derby televised by, 202-204
   experiments with human eye in television, 81-82

Baird, John L.—(*Continued*)
   images sent to S.S. *Berengaria* by, 84-85
   play televised by, 148-150
   television technique discussed by, 73-75
   transatlantic television test by, 79-81
   vaudeville demonstration by, 108
Bakewell, F. C., picture transmission process of, 6
Barkhausen, Dr. Heinrich, develops Barkhausen tube, 186-187
Barton, Bruce, discusses television's relation to print, 242-243, 260
Bell Telephone Laboratories, Ives describes television process of, 70-73
   observations of television by, 19-29
   outdoor television camera demonstrated by, 85-87
   television improvements by, 159-163
   television in color demonstrated by, 91-97
   two-way television on wire demonstrated by, 111-118
   Washington-New York television test by, 65-73
Berengaria, S.S., picks up television images at sea, 84-85
Beverage, Harold H., experiments with ultra-short waves, 184-186
Booths, developed for television speakers, 111
Braun tube, compared to cathode ray tube, 104
Broadcasting, early developments in, 135-138
   pioneer entertainers in, 210
   television's relation to, 238-241
   world-wide influence of, 208-209
Bureau of Standards, method to eliminate television "ghosts," 164-165

INDEX

Byrd, Rear Admiral R. E., discusses television's possibilities in exploration, 260-261

Caesium, use in photoelectric cells, 7, 22, 159-163
Caldwell, O. H., discusses ultra-short waves, 188-189
Camera, outdoor device for television, 85-87
See also Television camera
Carty, General J. J., participates in television test, 68-69
Cathode ray tube, Farnsworth's application of, 150-153
  first use in television, 55
  types of, 101-102
  Zworykin's use of, 97-104
Cell. *See* Photoelectric cell
Chamberlain, A. B., discusses art of television make-up, 169-172
Color television, correlated with music, 204-206
  Ives discusses, 95-96
  new tubes improve, 159-163
Commercial possibilities in television, 221-259
Conto, Armando, develops film device for television, 172-174
Crater tube. *See* Neon lamp
Crookes, Sir William, discovers cathode rays, 100
Crosley, Powel, comments on television, 195
Cutten, Dr. George B., television in education discussed by, 261-262

Dauvellier, Alexandre, experimenter in television, 55
Definition, by Baird, 73
  by Kennelly, 13-14
  television, 4
deForest, Dr. Lee, comments on television, 110, 195, 222-223
  future of television discussed by, 262-263
  three-element tube invented by, 8
Drama, television demonstration of, 87-90

Edison, Thomas Alva, "Edison effect" discovered by, 40-41
"Edison effect," discovery of, 40-41

Education, Cutten discusses television in, 261-262
  television possibilities in, 245-248, 251-259
Einstein, Dr. Albert, theory of radio transmission, 35-38
Ekstrom, early observations of scanning, 10
Electrical eye. *See* Photoelectric cell
Electrical scanning, advantages of, 97-105
  Farnsworth's experiments in, 150-153
  Zworykin's application of, 97-104
Electromagnetic waves, theories of, 34-41
Electronics. *See* Electrons
Electrons, action in cathode ray tube, 97-105, 151-153
  function in television, 19-29, 31-41
Element No. 87, possibilities for use in photoelectric cell, 22
Elster and Geitel, discover metals possess photoelectric properties, 7
Empire State Building, as television station site, 198-199
English Derby, televised for first time, 202-204
Ether, theories of, 34-41
Exploration, Byrd discusses television in, 260-261
  possibilities of television in, 245, 260-261
Eye, applied to television circuit, 81-82
  relation to television, 19-30, 73-74

Faraday, Michael, observations of electromagnetic waves, 36-37
Farnsworth, Philo T., contributions to television, 12
  experiments in electrical scanning, 150-153
Federal Radio Commission, report on television status, 168-169
Films, as used in television, 172-174
Fleming, John Ambrose, invents two-element valve, 9
Freeman, the Right Rev. James E., discusses television possibilities in religion, 263

INDEX

"Ghosts," in television, 48-49, 163-165
  method to exterminate, 164-165
Gifford, Walter S., comments on television, 67, 112-113
Goldsmith, Dr. Alfred N., looks ahead to 1940, 145-148
Gray, Dr. Frank, television camera described by, 86-87
  two-way television discussed by, 114
Griffith, D. W., participates in transcontinental television test, 109

Hammond, John Hays, Jr., contribution to research, 12
  invents television "eye" for aircraft, 127-134
Harbord, Major General James G., television application to war discussed by, 263-264
Harper's Weekly, prediction of television in 1900, 221
Hart, R. M., receives transatlantic television image, 79-81
Heaviside surface, influence on television, 163-165
  ultra-short waves affected by, 182-183
Hertz, Heinrich, experiments with wireless waves, 8
Hoover, Herbert C., comments on television, 69-70
  participates in television test, 65-70
Howe, Mrs. Mia, televised across Atlantic, 79-81
Hull, Ross A., television survey by, 197

Images, converting radio signals into, 113-116
  electrical scanning of, 97-105
  first to cross Atlantic, 79-81
  how formed in television, 18-30, 74-75
  how influenced by transmission, 47-49
  in color, 91-97
  in double form or "ghosts," 48-49, 163-165
  mechanical scanning of, 58-62, 70-73, 104-105, 113-116

Images—(*Continued*)
  picked up by S.S. *Berengaria,* 84-85
  projected on theater screen, 118-126
  sent to Australia and back, 106-108
  technique of tuning for, 148-150
  transmitted on light beam, 212-214
Infra-red rays, use in television, 75, 83-84
International relations, television's influence on, 225-226, 229-230, 251-253
International Telephone & Telegraph Company, ultra-short waves demonstration across English Channel, 177-180
Inventors in television, 10-13, 47-49, 55
Ives, Dr. Herbert E., Bell Telephone Laboratories' television described by, 70-73
  color television explained by, 95-97
  contributions to television, 12
  two-way television discussed by, 116-118

Jenkins, C. Francis, analyzes television, 52-54
  early work in radio-vision, 11-12
  lantern-slide scanning of, 214-216
Jewett, Dr. Frank B., comments on television status, 109-110

Karolus, Dr. August, contribution to television, 12-13
  light valve invented by, 120-121
Karplus, E., characteristics of ultra-short waves observed by, 187-189
Kennelly, Arthur E., defines television, 13-14
Kinescope, Zworykin introduces, 97-100
Knudsen, Hans, sends pictures by wireless, 10
Korn, Arthur, picture transmission by, 7

Lafount, Harold A., comments on television progress, 194-195

Lantern-slide scanning, as introduced by Jenkins, 214-216
Light, as handled in two-way television, 111-116
  images transmitted on beam of, 212-214
  relation to television, 18-30, 58-62
Lindbergh flight, as it might have been televised, 254-259
Lindenblad, Dr. N. E., experiments with ultra-short waves, 184-186

Make-up, technique in television, 149-150, 169-172
Manson, Ray H., comments on television, 196
Marconi, Guglielmo, comments on wireless-vision, 14-15, 49-50
  paves way for television, 15-17
Maxwell, James Clerk, editorial on ether theory of, 36-37
  theory of radio waves, 8
Mechanical scanning, advantages of, 104-105
  See also Scanning; Scanning disk
Metals, possessing photoelectric properties, 7
Microphone, how television camera supplements, 85-87
Micro-rays. *See* Ultra-short waves
Milhaly, Denoys von, experimenter in television, 55
Mills, John, explains television process, 19-29
Motion pictures, television's relation to, 230-234
Music, television's relation to, 234-236

National Advisory Council on Radio in Education, reports on value of television in education, 246-247
Neon lamp, compared with argon lamp, 96-97
  function in television, 24-27, 153-156
  improvements in, 161-163
News events, television application to, 251-259

New York Evening Post, comment on televising English Derby, 202-203
New York Times, Maxwell's ether theory (editorial), 36-37
  television's future (editorial) 219-220
  television's progress (editorial) 45-46
Nipkow, Paul, invents television scanning disk, 10-11
Noctovisor, development of, 83-84

Olpin, A. R., work in television research, 159-163
Opera, television's influence on, 235-236
Optical nerve, television extension of, 18-30

Paley, William S., outlines television status, 194
Parkin, Sir George R., comments on Marconi transoceanic demonstration, 15-17
Patterson, E. P., color music discussed by, 204-206
Persistence of vision, in television, 27-30
Photoelectric, derivation of word, 4
  metals possessing properties, 7
  See also Photoelectric cell
Photoelectric cell, as television eye, 4-5
  Bell Telephone Laboratories' use of, 70-73
  caesium type of, 159-163
  color television use of, 91-97
  how it functions, 19-30
  metals used in, 7, 22
  Stoletow's work with, 7
  television booth's arrangement of, 114-116
  two-way television use of, 111-118
Photoradio, Ranger's system, 8-10, 50-52
  transatlantic demonstration, 52
Picture transmission, Amstutz sends a half-tone, 7
  Bain's experiments, 5-6
  Bakewell's process, 6
  Jenkins' experiments in, 11-12
  Korn's early attempt at, 7
  pioneers in, 7

INDEX

Picture transmission—(*Continued*)
Ranger's system, 8-10, 50-52
selenium used in, 6-7
transatlantic demonstration of, 52
Picturegram. *See* Photoradio
Police, television possibilities, 244-245
Politics, Roosevelt discusses television in, 264-265
television as an influence in, 125-126, 236-238, 264-265
Potassium, use in photoelectric cell, 7
Print, television's relation to, 241-244
Programs, examples of W2XAB, 210-212, 227-228

Quasi-optical waves. *See* Ultra-short waves.

Radio, Einstein's theory of, 35-38
Marconi's early experiments in, 14-17
Maxwell theory of waves, 8, 36-37
pictures sent by, 8-10, 50-52
relation to television, 31-41
Steinmetz theory of, 35, 38-40
Radio City, plans for, 134-144
Radio Corporation of America, observations of ultra-short waves, 182-186
statement on status of television, 165-166
Radio waves, Einstein and Steinmetz theories of, 35-40
Hertz produces and detects, 8
Maxwell theory of, 8, 36-38
Ranger, Capt. Richard, photoradio system of, 8-10, 50-52
Reis, Philip, observations of selenium, 6
Religion, television's influence on, 263
Retina, action of television upon, 27-30, 73-74
Rockefeller Center. *See* Radio City
Roosevelt, Colonel Theodore, discusses television's influence on politics, 264-265
Rosing, Boris, experiments in television, 55

Rothafel, Samuel L. (Roxy), discusses television's relation to stage and screen, 265

Sanabria, Ulisses A., contribution to television, 13
demonstrates his system, 206-207
observation of television "ghosts," 164-165
Sarnoff, David, Radio City possibilities discussed by, 138-139
television problems outlined by, 193-194
television's relation to motion pictures discussed by, 233-234
Scanning, Alexanderson's analysis of, 58-62
Baird's explanation of, 29-30, 55-57
electrical method of, 97-105, 150-153
in two-way television system, 113-116
Ives explanation of, 70-73
lantern-slide method of, 214-216
mechanical method of, 58-62, 104-105
outdoor scenes, 85-87
Scanning disk, Baird's description of, 55-57
function of, 24-30
invention of, 10-11
Ives describes action of, 70-73
neon lamp's relation to, 153-156
use with outdoor camera, 86
Selenium, use in picture transmission, 6-7
Smith, Willoughby, investigation of selenium, 6
Sports, television possibilities in, 227, 232-234
Static, effect on television, 166-168
Steinmetz, Charles Proteus, theory of radio waves, 35, 38-40
Stoletow, makes photoelectric cell, 7
Super-regenerator, use in ultra-short wave reception, 188-189

Telecast. *See* Television.
Telephony, relation to television, 20-22
television's influence on, 226

296 INDEX

Telephony—(*Continued*)
two-way television demonstration, 111-118
Televise. *See* Television
Television, advertising by, 227-229
aircraft applications of, 125, 244-245
Alexanderson describes his system, 58-62
Alexanderson predicts future of, 62-64, 88-89, 122-126
Aylesworth outlines possibilities of, 140-141, 199-201, 240-241
Baird defines, 73
Baird describes his system, 55-57
Baird's method of approach to, 73-75, 82-84
Bell Telephone Laboratories' color method in, 91-97
Bell Telephone Laboratories' demonstration between Washington and New York, 65-73
Bell Telephone Laboratories demonstrate outdoor camera, 85-87
Bell Telephone Laboratories' improvements in, 159-163
Bell Telephone Laboratories' two-way system of, 111-118
Berengaria intercepts images, 84-85
cathode-ray tubes in, 97-105
color in, 91-97
color-music in, 204-206
commercial destiny of, 221-259
contemporary inventors in, 10-13
definition of, 4, 73
DeForest comments on, 110, 195, 222-223, 262-263
derivation of the word, 3-4
educational possibilities of, 245-248, 261-262
electrical scanning in, 97-105
English Derby televised, 202-204
experiment with human eye in, 81-82
explorers' use of, 245-248, 260-261
"eye" for airplanes, 127-134
Farnsworth's experiments in, 150-153
films used in, 172-174
first one-act play by, 87-90
first transatlantic test of, 79-81
first vaudeville act by, 66

Television—(*Continued*)
"ghosts" in, 48-49, 163-165
Gifford's comments on, 67
Goldsmith looks ahead to, 145-148
Hammond's "eye" for aircraft, 127-134
Hoover comments on, 69-70
Hoover participates in test, 65-70
how a person is televised, 111-112
human eye's relation to, 19-30
images transmitted on light beam, 212-214
incentives for experimenters, 124
international relations influenced by, 225-226, 229-230
Ives describes Bell Laboratories' process of, 70-73
Jenkins' analysis of, 52-54
Jenkins' lantern-slide method of, 214-216
Jenkins' prediction of, 11-12
Kennelly's definition of, 13-14
lantern-slide scanning in, 214-216
limitations of, 197, 199-201
make-up in, 150-153, 168-173
Marconi's comments on, 14-17
mechanical scanning process in, 29-30
Mills' explanation of, 19-29
motion-picture's relation to, 230-234, 265
music's relation to, 234-236
names suggested for audience, 174-175
neon lamp in, 24-30, 153-156
news events handled by, 251-259
obstacles foreseen, 222-224
opera possibilities in, 236
outdoor camera demonstration, 85-87
Paley discusses possibilities of, 194
police application of, 244-245
political possibilities in, 125-126, 236-238, 264-265
possibilities in exploration, 245, 260-261
print's relation to, 241-244, 260
process simply described, 166-168
progress observed by R C A, 165-166
Radio City's plans for, 134-144
religion as influenced by, 263

INDEX

Television—(*Continued*)
 roster of stations, 288-289
 Sarnoff discusses problems in, 193-194
 scanning-disk invented for, 10-11
 Schenectady-Australia test of, 106-108
 scientific status of, 108-111, 192-198
 sound broadcasting relation to, 238-241
 sporting events sent by, 227, 232-234
 technique of tuning, 148-150
 telephony's relation to, 20-22, 226
 theater's relation to, 138-144, 230-234, 265
 theater-screen demonstration of, 118-122
 The New York Times' editorial on progress of, 45-46, 219-220
 time sent by, 226-227
 transatlantic test of, 79-81
 ultra-short waves in, 177-191
 vacuum tube's function in, 8, 31-34
 vaudeville by, 108
 war uses of, 133-134, 244, 263-264
 Washington-New York test of, 65-73
 what Hammond foresees in, 127-134
 world-wide performance of, 251-259
 W2XAB's programs, 210-212
 W2XCR's première, 166-168
 Zworykin's system of, 97-104
Television camera, art of focusing, 169-172
 Dr. Frank Gray discusses, 86-87
 supplements microphone, 85-87
 See also Camera
Television "eye," Hammond's development of, 127-134
 See also Photoelectric cell
Television projector, developed for theater use, 118-122
Theater, Rothafel discusses television's relation to, 265
 Sarnoff discusses television's relation to, 230-234
 television demonstrated in, 118-122

Theater—(*Continued*)
 television's relation to, 138-144, 230-234
Thomson, Elihu, predicts televising sun's eclipse, 201
Time, sent by television, 226-227
Transatlantic, photoradio demonstration, 52
 television test, 79-81

Ultra-short waves, applications of, 183-184
 Baird's observations of, 190
 characteristics of, 187-191
 images sent on light beam, 212-214
 possibilities outlined by Caldwell, 189-190
 test across English Channel, 177-180
 tests in New York, 183-186
 use in Hawaii, 180-181, 190-191
 use in television, 176-191
 Warner discusses possibilities, 181-182

Vacuum tube, DeForest's invention, 8
 Edison's contribution to, 40-41
 Fleming's invention, 8
 function in television, 8, 31-41
Van Hoogstraten, Willem, discusses television's relation to music, 234-235

Wade, Clem F., report on television progress, 194
War, Harbord foresees television in, 263-264
 use of television in, 133-134, 244, 263-264
Warner, Kenneth B., ultra-short wave possibilities discussed by, 181-182
W2XAB, experiments in television make-up, 168-173
 programs televised by, 210-212, 227-228
W2XCR, première performance of, 166-168

Zworykin, Vladimir, contribution to television, 12
 Kinescope introduced by, 97-100
 progress by, 201

HISTORY OF BROADCASTING:
Radio To Television
An Arno Press/New York Times Collection

Archer, Gleason L.
Big Business and Radio. 1939.

Archer, Gleason L.
History of Radio to 1926. 1938.

Arnheim, Rudolf.
Radio. 1936.

Blacklisting: Two Key Documents. 1952–1956.

Cantril, Hadley and Gordon W. Allport.
The Psychology of Radio. 1935.

Codel, Martin, editor.
Radio and Its Future. 1930.

Cooper, Isabella M.
Bibliography on Educational Broadcasting. 1942.

Dinsdale, Alfred.
First Principles of Television. 1932.

Dunlap, Orrin E., Jr.
Marconi: The Man and His Wireless. 1938.

Dunlap, Orrin E., Jr.
The Outlook for Television. 1932.

Fahie, J. J.
A History of Wireless Telegraphy. 1901.

Federal Communications Commission.
Annual Reports of the Federal Communications Commission. 1934/1935–1955.

Federal Radio Commission.
Annual Reports of the Federal Radio Commission. 1927–1933.

Frost, S. E., Jr.
Education's Own Stations. 1937.

Grandin, Thomas.
The Political Use of the Radio. 1939.

Harlow, Alvin.
Old Wires and New Waves. 1936.

Hettinger, Herman S.
A Decade of Radio Advertising. 1933.

Huth, Arno.
Radio Today: The Present State of Broadcasting. 1942.

Jome, Hiram L.
Economics of the Radio Industry. 1925.

Lazarsfeld, Paul F.
Radio and the Printed Page. 1940.

Lumley, Frederick H.
Measurement in Radio. 1934.

Maclaurin, W. Rupert.
Invention and Innovation in the Radio Industry. 1949.

Radio: Selected A.A.P.S.S. Surveys. 1929–1941.

Rose, Cornelia B., Jr.
National Policy for Radio Broadcasting. 1940.

Rothafel, Samuel L. and Raymond Francis Yates.
Broadcasting: Its New Day. 1925.

Schubert, Paul.
The Electric Word: The Rise of Radio. 1928.

Studies in the Control of Radio: Nos. 1–6. 1940–1948.

Summers, Harrison B., editor.
Radio Censorship. 1939.

Summers, Harrison B., editor.
**A Thirty-Year History of Programs Carried on
National Radio Networks in the United States, 1926–1956.** 1958.

Waldrop, Frank C. and Joseph Borkin.
Television: A Struggle for Power. 1938.

White, Llewellyn.
The American Radio. 1947.

World Broadcast Advertising: Four Reports. 1930–1932.